MARS

CONTENTS

I.

ATMOSPHERE.

AMID the seemingly countless stars that on a clear night spangle the vast dome overhead, there appeared last autumn to be a new-comer, a very large and ruddy one, that rose at sunset through the haze about the eastern horizon. That star was the planet Mars, so conspicuous when in such position as often to be taken for a portent. Large as he then looked, however, he is in truth but a secondary planet traveling round a secondary sun; but his interest for us is out of all proportion to his actual size or his relative importance in the cosmos. For that sun is our own; and that planet is, with the exception of the moon, our next to nearest neighbor in space, Venus alone ever approaching us closer. From him, therefore, of all the heavenly bodies, may we expect first to learn something beyond celestial mechanics, beyond even celestial chemistry; something in answer to the mute query that man instinctively makes as he gazes at the stars, whether there be life in worlds other than his own.

Hitherto the question has received no affirmative reply, although the trend of all latter-day investigation has been to such affirmation; for science has been demonstrating more and more clearly the essential oneness of the universe. Matter proves to be common property. We have learnt that the very same substances with which we are familiar on this our earth, iron, magnesium, calcium, and the rest, are present in the far-off stars that strew the depths of space. Nothing new under the sun! Indeed, there is nothing new above it but ever-varying detail.

So much for matter. As for mind beyond the confines of our tiny globe, modesty, backed by a probability little short of demonstration, forbids the thought that we are the sole thinkers in this great universe.

That we are the only minds in space it takes indeed a very small mind to fancy. Our relative insignificance commonly escapes us. If we reduce the universe to a scale on which we can conceive it, that on which the earth shall be represented by a good-sized pea, with a grain of mustard seed, the moon, circling about it at a distance of seven inches, the sun would be a globe two feet in diameter, two hundred and twenty feet away. Mars, a much smaller pea, would circle round the two-foot globe, three hundred and fifty feet from its surface; Jupiter, an orange, at a distance of a quarter of a mile; Saturn, a small orange, at two fifths of a mile; and Uranus and Neptune, good-sized plums, three quarters of a mile and a mile and a quarter away, respectively. The nearest star would lie two hundred and thirty thousand miles off, or at about the actual distance of our own moon, and the other stars at corresponding distances beyond that; that is, on a scale upon which the moon should be but seven inches off, the nearest star would still be as far from us as the moon is now. When we think that each of these stars is probably the centre of a solar system on a grander scale than our own, we cannot seriously take ourselves to be the only minds in the universe.

But improbable as the absence of ultra-terrestrial life in a general way is, up to the present time we have had no proof of its particular existence in worlds beyond our own. Whether the observations I am now to describe have revealed something on the point I shall leave the reader himself to judge, after laying the facts before him; for it is with this in view that the present papers will deal with Mars, since any answer on this point is the most generally interesting outcome of a study of the planet. That the observations also disclose the fact that the hitherto accepted period of its rotation proves to be too small by the hundredth of a second is a matter of far greater moment, of course, but one which leaves the average man comparatively cool. That Mars, however, should be peopled by intelligent beings, although physically they be utterly unlike us, more goblins than men or animals, is a suggestion which appeals romantically, at least, to everybody.

To determine whether a planet be the abode of life, two questions about it must be answered in turn: first, are its physical conditions such as to render it habitable? and secondly, are there any signs of its actual habitation? Unless we can answer the first point satisfactorily, it were futile to seek for evidence of the second.

Of such planets as doubtless circle round other suns we as yet know nothing. Our search is perforce confined at present to the members of our own solar family. Now, when we scan them for answer to our first query, we find but two that promise even the possibility of an affirmative reply, Mars and Venus. All the others turn out, upon scrutiny, to lie beyond the pale, either because they are too big, or because they are too little; for, curiously enough, mere size settles the matter.

The giant Jupiter piques inquiry first by showing us great cloud-belts that recall our own equatorial and temperate cloud-zones. But further study discloses that his clouds are in kind quite unlike those of our earth. Neither the hour of his day nor the season of his year brings changes in them. They slowly, very slowly, alter in appearance, indeed, but not in obedience to that central ruler that gathers and dispels our own. In short, the Jovian clouds are not sun-raised, but self-raised ones. It is heat inherent in Jupiter himself, not heat from the sun, that belts him about with his great girdles of cloud. We can even see, in all probability, his glowing inner self; for Jupiter shows brick-red between his belts, like a molten mass.

The same state of things is yet more strikingly instanced by Saturn; for the tilt of Saturn's pole is not very unlike that of the earth, and in consequence his equatorial regions are at times raised far above the plane of his orbit; at others, dipped far below it. Yet unlike the earth's cloud-belts, his never travel northward when the sun goes north, nor follow the sun when he journeys south again. So far as the sun is concerned, the Saturnian cloud-belts are invariable. Like the Jovian, they owe their formation to the planet's own heat. Like Jupiter, too, Saturn shows red beneath.[1] From all this it is pretty plain that the giant planets are far from pleasurable abodes, as yet midway in evolution between actual suns and tenantable worlds; too cooled down for the one state, and not yet cooled down enough for the other.

Uranus and Neptune give evidence, also of being in a chaotic condition, orbs *informe, ingens, cui lumen ademptum,*—no longer suns, but as yet quite unfit to support beings even distantly analogous to ourselves.

With Mercury littleness is even more fatal to life; for though the giant planets may perhaps, at some future day, grow to be life-supporting, a small one apparently never was, nor ever can be, peopled by beings in the least resembling us. Incapacity to quarter folk is included in the more general incapacity to hold an atmosphere; for absence of atmosphere precludes the possibility of life as we know it. That a planet may be too small to have an atmospheric envelope we shall see more definitely later. That life, however, of a type of which we have no conception may not exist in all these orbs we must be wary of stating, for nothing is more dangerous than a general denial, except a particular statement.

We are limited, therefore, in our present inquiry, to Venus and Mars. But Venus, contrary to her name, proves provokingly modest, the most modest of all the company of heaven, keeping herself so constantly veiled in cloud that we seldom, if ever, are permitted a peep at her actual surface. In consequence, beyond the fact that she has an atmosphere of considerable though not excessive density, we know little about her.

---

1   Both Jupiter and Saturn are ruddier than is commonly stated. In the air of Flagstaff, Arizona, the site of my observations, both of them show conspicuously red between their belts.

With Mars, on the other hand, no such false modesty balks us at the outset. The planet named after the old God of War—satirically, it would seem, since he turns out to present characteristics quite the reverse of warlike—lets himself be seen as well as thirty-five millions of miles of separation will allow.

Now, to all forms of life of which we have any conception, two things in nature are vital, air and water. A planet must possess these two things to be able to support any life at all upon its surface. Some articles that we might deem essential to well-being fall cosmically under the head of luxuries; but air and water are necessities of existence. There is no creature which is not in some measure dependent upon both of them. How then is Mars off for air?

Fortunately for an answer to this question, air is as vital to change in the inorganic processes of nature as it is to those other changes which we call peculiarly life. Atmosphere is essential not only to life upon a planet, but to the production of any change whatever upon that planet's surface. Without it, not only development, but decay would come to a standstill, when once all that was friable had crumbled to pieces under the alternate roasting and refrigerating to which the planet's surface would be exposed as it revolved upon its axis toward and away from the sun. Disintegration once effected, the planet would roll, a mummy world, through space. Since atmosphere, therefore, is a *sine qua non* to any change upon a planet's surface, reversely, any change upon a planet's surface is proof positive of the presence of an atmosphere, however incapable of detection such atmosphere be by direct means.

Now changes take place upon the surface of Mars, changes vast enough to be visible from the earth. When properly observed they turn out to be most marked. We will begin with the look of the planet last June. Its general aspect then was tripartite. Upon the top part of the disc, round what we know to be the planet's pole, appeared a great white cap, the south polar cap. The south lay at the top, because all astronomical views are, for optical reasons, upside down; but inasmuch as we never see the features otherwise, to have them right side up is not vital to the effect. Below the white cap lay a region chiefly bluish-green, interspersed, however, with portions more or less reddish-ochre. Below this, again, came a vast reddish-ochre stretch, the great continental deserts of the planet.

The first sign of change occurred in the polar cap. It proceeded slowly to dwindle in size. Such obliteration it has, with praiseworthy regularity, undergone once every two years for the last two hundred. Since the polar cap was first seen it has waxed and waned with clock-like precision, a precision timed to the change of season in the planet's year. During the spring, these snow-fields, as analogy at once guesses them to be, and as beyond doubt they really are, stretch in the southern hemisphere, the one presented to us at this last opposition, down to latitude seventy, and even sixty-five south; covering thus more than the whole of the planet's south frigid zone. As summer comes on they dwindle gradually away, till by early autumn they present but tiny patches, a few hundred miles across. This year, for the first time in human experience, they melted, apparently, completely. This unprecedented event happened on October 13, or forty-three days after the summer solstice of the southern hemisphere, a date corresponding to about the middle of July on earth. Evidently it was a phenomenally hot season on Mars, for the minimum of the polar patch is reached usually about three months after Martian midsummer. It will be noticed how nearly such melting parallels what takes place with our arctic ice-cap on earth.

But the disappearance of the polar snows is by no means the only change discernible upon the surface of the planet. Several years ago Schiaparelli noticed differences in tint at successive oppositions, both in the dark areas and in the bright ones. These, he suggested, might be due to the seasons. This year it has been possible to watch the change take place. From the Martian last of April to the Martian middle of August, the bluish-green areas have been steadily undergoing a most marked transformation. There proves, in fact, to be a wave of seasonal

14

change that sweeps over the face of the planet from pole to pole. We will examine this more in detail when we take up the question of water. For the present point it suffices that it takes place; for it constitutes proof positive of the presence of an atmosphere.

A moment's consideration will show how absolutely positive this proof is; for it is the inevitable deduction from the simplest of observed facts. Its cogency consists in its simplicity. It is independent of difficult detail or of doubtful interpretation. It is not concerned with what may be the constitution of the polar caps, nor with the character of the transformation that sweeps, wavelike, over the rest of the planet. It merely states that change occurs, and that statement is conclusive.

Having thus seen with the brain as much as with the eye, and in the simplest possible manner, that a Martian atmosphere exists, we will go on to consider what it is like.

The first and most conspicuous of its characteristics is its cloudlessness. A cloud is an event on Mars, a rare and unusual phenomenon, which should make it more fittingly appreciated there than Ruskin lamented was the case on earth. For it is almost perpetually fine weather on our neighbor in space. From the day's beginning to its close, and from one end of a year to the other, nothing appears to veil the greater part of the planet's surface.

This is more completely the case than has hitherto been supposed. We read sometimes in astronomical books and articles picturesque accounts of clouds and mist gathering over certain regions of the disc, hiding the coast lines and continents from view, and then, some hours later, clearing off again. No instance of such blotting out of detail has been seen this year at Flagstaff. Though the planet's face has been scanned there almost every night, from the last day of May to the end of November, not a case of obscuration of any part of the central portions of the planet, from any Martian cause, has been detected by any one of three observers. Certain peculiar brightish patches have from time to time been noted, but, with a courtesy uncommon in clouds, they have carefully refrained from obscuring in the slightest degree any detail the observer might be engaged in looking at.

The only dimming of detail upon the Martian disc has been along its bright edge, what is technically called its limb. Fringing this is a permanent lune of light that swamps all except the very darkest markings in its glare. This limb-light has commonly been taken as evidence of sunrise or sunset mists on Mars. But observations of mine during last June show that such cannot be the case. In June Mars was gibbous,—that is, he showed a face like the moon between the quarter and the full,—and along his limb, then upon his own western side, lay the bright limb-light, stretching inward about thirty degrees. Since the face turned toward us was only in part illumined by the sun, the centre of it did not stand at noon, but some hours later, and the middle of the limb consequently not at sunrise, but at about nine o'clock of a Martian morning. As the limb-light extended in from this thirty degrees, or two hours in time, the mist, if mist it was, must have lasted till eleven o'clock in the day. Furthermore, it must have been mist of a singularly mathematical turn of mind, for it made a perfect semi-ellipse from one pole to the other, quite oblivious of the fact that every hour from sunrise to sunset lay represented along its edge, including high noon. What is more, as the disc passed, in course of time, from the gibbous form on the other side, the limb-light obligingly clung to the limb, regardless of everything except its geometric curve. But as it did so, the eleven o'clock meridian swung from one side of the centre of the disc to the other. As it crossed the centre its regions showed perfectly clear; not a trace of obscuration as it passed directly under the eye. It was evident, therefore, that Martian morning mists were not responsible for the phenomenon.

To what, then, was the limb-light due? At first sight, it would seem as if the moon might help us; for the moon's limb is similarly ringed by a lune of light. In her case the effect has been attributed to mountain slopes catching the sun's light at angles beyond the possibilities of plains. But Mars has few mountains worthy the name. His terminator—that

16

is, the part of the disc which is just passing in or out of sunlight, and discloses mountains by the way in which they catch the coming light before the plains at their feet are illuminated—shows irregularities quite inferior to the lunar ones, proving that his elevations and depressions are relatively insignificant.

On the whole, the best explanation of the phenomenon seems to be that the Martian atmosphere itself is somewhat of a veil, and that this veiling effect, though practically imperceptible in the centre of the disc, becomes noticeable as we go from the centre to the edge, owing to the greater thickness of the stratum through which we look. At thirty degrees in from the limb the observer would look through twice as much of it as when he looked plumb down upon the centre of the disc; in consequence, what would be diaphanous at the centre might well seem opaque toward the edge. The effect we are familiar with on earth in the haze that always borders the horizon,—a haze most noticeable in places where there is much water in the air. Here, then, we have a hint of what is the matter on Mars. Were his atmosphere charged with water-vapor, just such an effect as is observed should take place.

This first hint receives independent support from another Martian phenomenon. Contrary to what the distance of the planet from the sun and the thinness of its atmospheric envelope would lead us to expect, the climate of Mars proves astonishingly mild. Whereas calculation from distance and atmospheric density puts its average temperature below freezing, thus relegating it to perpetual ice, the planet's surface features show that the temperature is relatively high. Observation reveals the fact that the mean temperature must actually be above that of the earth; for not only is there practically no snow or ice outside the frigid zone at any time, but the polar snow-caps melt to a minimum quite beyond that of our own, affording the Martians rare chance for quixotic polar expeditions. Such pleasing amelioration of the climate must be accounted for, and aqueous vapor seems the most likely thing for the purpose; for aqueous vapor is quite specific as a planetary comforter, being the very best of blankets. It acts, indeed, like the glass of a conservatory, letting the light rays in, and opposing the passage of the heat rays out.

The state of things thus disclosed by observation, the cloudlessness and the rim of limb-light, turns out to agree in a most happy manner with what probability would lead us to expect; for the most natural supposition to make *a priori* about the Martian atmosphere is the following. When each planet was produced by fission from the parent nebula, we may suppose that it took with it as its birthright its proportion of chemical constituents; that is, that its amount of oxygen, nitrogen, and so forth was proportional to its mass. Doubtless its place in the primal nebula would to a certain extent modify the ratio, just as the size of the planet would to a certain extent modify the relative amount of these elements that would thereupon enter into combination. Supposing, however, that the ratio of free oxygen and so forth to the other elements remained substantially the same, we should have in the case of any two planets the same relative quantity of atmosphere. But the size of the planet would entirely alter the distribution of this air.

Three causes would all combine to rob the smaller planet of efficient covering, on the general principle that he that hath little shall have less.

In the first place, the smaller the planet, the greater would be its volume in proportion to its mass, because the materials of which it was composed, being subjected to less pressure owing to a lesser pull, would not be crowded so closely together. This is one reason why Mars should have a thinner atmosphere than is the case with our earth.

Secondly, of two similar bodies, spheres or others, the smaller has the greater surface for its volume, since the one quantity is of two dimensions only, the other of three. An onion will give us a good instance of this. By stripping off layer after layer we reach eventually a last layer which is all surface, inclosing nothing. We may, if we please, observe something analogous in men, among whom the most superficial have the least in them. In consequence of this principle, the atmosphere of the smaller body finds itself obliged to cover relatively more surface, which still further thins it out.

Lastly, gravity being less on the surface of the smaller body, the atmosphere is less compressed, and, being a gas, seizes that opportunity to spread out to a greater height, which renders it still less dense at the planet's surface.

Thus for three reasons Mars should have a thinner air at his surface than is found on the surface of the earth.

Calculating the effect of the above causes numerically, we find that on this a priori supposition Mars would have at his surface an atmosphere of about fourteen hundredths, or one seventh the density of our terrestrial one.

Observation supports this general supposition; for the cloudless character of the Martian skies is precisely what we should look for in a rare air. Clouds are congeries of globules of water or particles of ice buoyed up by the air about them. The smaller these are, the more easily are they buoyed up, because gravity, which tends to pull them down, acts upon their mass, while the resistance they oppose to it varies as their surface, and this, as we saw just now, is relatively greater in the smaller particles. The result is that the smaller particles can float in thinner air. We see the principle exemplified in our terrestrial clouds; the low nimbus being formed of comparatively large globules, while the high cirrus is made up of very minute particles. If we go yet higher, we reach a region incapable of supporting clouds of any kind, so rarefied is its air. This occurs about five miles above the earth's surface; and yet even at this height the density of our air is greater than is the probable density of the air at the surface of Mars. We see, therefore, that the Martian atmosphere should from its rarity prove cloudless, just as we observe it to be.

So far in this our investigation of the Martian atmosphere we have been indebted solely to the principles of mathematics and molar physics for help, and these have told us something about the probable quantity of that atmosphere, though silent as to its possible quality. On this latter point, however, molecular physics turns out to have something to say; for an Irish gentleman, Dr. G. Johnstone Stoney, has recently made an ingenious deduction from the kinetic theory of gases bearing upon the atmosphere envelope which any planet can retain. His deduction is as acute as it appears from observation to be in keeping with the facts. It is this:—

The molecular theory of gases supposes them to be made up of myriads of molecules in incessant motion. What a molecule may be nobody knows; some scientists supposing it to be a vortex ring in miniature,— something like the swirl produced by a teaspoon when drawn through a cup of tea. But whatever it be, the idea of it accounts for the facts. The motion of the molecules is almost inconceivably swift as they dart hither and thither throughout the space occupied by the gas, and their speed differs for different gases. It is calculated that the molecules of oxygen travel, on the average, at the rate of fifteen miles a minute, those of water vapor about twenty miles a minute, and those of hydrogen, which are the fastest known, at the enormous speed of a mile a second. But this average velocity may, in any particular case, be increased by collisions of the molecules among themselves something like sevenfold. What is more, each molecule of the gas is bound, sooner or later, to attain this maximum velocity of its kind merely on the doctrine of chances. When it is attained, the molecule of oxygen travels at the rate of one and three fourths miles a second, the molecule of water vapor at the rate of two and one third miles a second, and the molecule of hydrogen actually at seven miles a second, six hundred times as fast as our fastest express train.

Now, if a body, whether it be a molecule or a cannon-ball, be projected away from the earth's surface, the earth will at once try to pull it down again: this instinctive holding on of Mother Earth to what she has we call gravity. In the cases with which we are personally familiar, her endeavor is eminently successful; what goes up usually coming down again, either on the thrower or on some other person. But even the earth is not omnipotent.

As the velocity with which the body is projected increases, it takes the earth longer and longer to overcome it and compel the body's return. Finally there comes a speed which the earth is just able to overcome, if she take an infinite time about it. In that case, the body would continue to travel away from her, at a constantly diminishing rate, but still at some rate, on and on into the depths of space, till it attained infinity, at which point the truant would stop, and reluctantly begin to return again. This velocity we may call the critical velocity. It is the velocity which the earth would cause in a body falling to it from an infinite distance, since gravity is able to destroy on the way up just the speed it is able to create on the way down. But now, if the body's departure were even hastier than this, the earth would never be able wholly to annihilate its speed, and the body would travel forever away out and out, till it fell, probably, under the sway of some distant star. In any case, the earth would know the vagabond no more.

As gravity depends upon mass, the larger the attracting planet, the greater is its critical velocity, the velocity it can just control; and, reversely, the smaller the planet, the less its restraining power. With the earth the critical velocity is between six and seven miles a second. If any of us, therefore, could manage to become faster than this, socially or otherwise, we could bid defiance to the whole earth, and begin to voyage on our own account through space.

This is actually what happens, as we have seen, to the molecules of hydrogen. If, therefore, free hydrogen were present at the surface of the earth, and met with no other gas attractive enough to tie it down by uniting with it, the rover would, in course of time, attain a speed sufficient to allow it to bid good-by to earth, and start on interspacial travels of its own. That it should reach its maximum speed is all that is essential to liberty, the direction of its motion being immaterial. To each molecule in turn would come this happy dispatch, till the earth stood deprived of every atom of free hydrogen she possessed.

It is a highly significant fact that there is no free hydrogen found in the earth's atmosphere. With oxygen and water vapor, and indeed all the other gases we know, the case is different; for their maximum speed falls far short of the possibility of escape. So they have stayed with us solely because they must. And, as a matter of fact, the earth's atmosphere contains plenty of free oxygen, nitrogen, and the like. The actions of the heavenly bodies confirm this conclusion. The moon, for example, possesses no atmosphere, and calculation shows that the velocity it can control falls short of the maximum of any of these gases. All were, therefore, at liberty to leave it, and all have promptly done so. Whatever the moon's attraction for lovers, no gas was sufficiently attracted by it to stay. On the other hand, the giant planets give evidence of very dense atmospheres. They have kept all they ever had.

But the most striking confirmation of the theory comes from the cusps of Venus and Mercury; for an atmosphere would prolong, by its refraction, the cusps of a crescent beyond their true limits. Length of cusp becomes, consequently, a criterion of the presence of an atmosphere. Now, in the appearance of their cusps there is a notable difference between Venus and Mercury. The cusps of Venus extend beyond the semi-circle; Mercury's do not. We see, therefore, that Mercury has no appreciable atmospheric envelope.

Turning to the case of Mars, we find with him the critical velocity to be about three miles a second. This is, like the earth's, below the maximum for the molecules of hydrogen, but also, like the earth's, above that of any other gas; from which we have reason to suppose that, except for possible chemical combinations, his atmosphere is in quality not unlike our own.

Having seen what the atmosphere of Mars is probably like, we may draw certain interesting inferences from it as to its capabilities for making life comfortable. The first consequence is that Mars is blissfully destitute of weather. Unlike New England, which has more than it can accommodate, Mars has none of the article. What takes its place as the staple topic of conversation for empty-headed folk there remains one of the Martian mysteries yet to be solved. What takes its place in fact is a perpetual serenity, such as we can scarcely conceive of. Although over what we shall later see to be the great continental deserts the air must at midday be highly rarefied, and cause vacuums into which the surrounding air must rush, the actual difference of gradient owing to the initial thinness of the air must be very slight. With a normal barometer of four and a half inches, a very great relative fall is a very slight actual one. In consequence, storms would be such mild-mannered things that, for objectionable purposes, they might as well not be. In the first place, if we are right, there can be no rain, nor hail, nor snow in them, for the particles would be deposited before they gained the dignity of such separate existence. Dew or frost would be the maximum of precipitation that Mars could support. The polar snow-cap or ice-cap, therefore, is doubtless formed, not by the falling of snow, but by successive depositions of dew. Secondly, there would be about the Martian storms no very palpable wind. Though the gale might blow at fairly respectable rates, so flimsy is the substance moved that it might buffet a man unmercifully without reproach.

Another interesting result of the rarity of the air would be its effect upon the boiling-point of water. Reynault's experiments have shown that, in air at a density of $14/100$ of our own, water would boil at about $127°$ Fahrenheit. This, then, would be the temperature at which water would be converted into steam on Mars. So low a boiling-point would make it impossible to cook anything in the open air. Boiled eggs could be prepared only under cover, and such people as liked their meat boiled would probably find it convenient to prefer it done differently. Fortunately, roasts would still remain possible. The lowering of the boiling-point would raise the relative amount of aqueous vapor held in suspension by the air at any temperature. At about $127°$ the air would be saturated, and even at lower temperatures much more of it would evaporate and load the surrounding air than happens at similar temperatures on earth. Thus at the heels of similarity treads contrast.

We may now go on to such phenomena bearing on the Martian atmosphere as show it to differ from ours. Some of them we are able more or less imperfectly to explain; some we are not.

Although no case of obscuration has been seen at Flagstaff this summer, certain bright patches have been observed on special portions of the planet's disc. That they are not storm-clouds, like those which, by a wavelike process of generation, travel across the American continent, for example, is shown by the fact that they do not travel, but are local fixtures. Commonly, they appear day after day, and even year after year, in the same spots; for identical patches have been observed by different astronomers at successive oppositions. To this category belong the regions known as Elysium, Ophir, Memnonia, Eridania, and Tempe. Still smaller patches, apparently more fugitive in character, have been seen this year by Professor W. H. Pickering. But the most marked instance of variability was detected in September last by Mr. Douglass, in the western part of Elysium. On September 22 and 23 he found this blissfully named region, as usual, equally bright throughout. But on September 24 he noticed that the western half of it had suddenly increased in brightness, and far outshone the eastern half, being almost as brilliant as the polar cap. When he looked at it again the next night, September 25, the effect of the night before had vanished, the western half being now actually the darker of the two. So fugitive an effect suggests cloud, forming presumably over high ground, and subsequently dissipating; it also suggests a deposition of frost that melted on the next day. It is specially noteworthy that the canals inclosing the region, Galaxias and Hyblaeus, were not in any way obscured by the bright apparition. On the contrary, Mr. Douglass found them perceptibly darker than they had been, an effect attributable perhaps to contrast.

Although not storm-clouds, it is possible that these appearances may have been due to cloud capping high land. There are objections, however, to this view, as, in the first place, there is evidence that the Martian mountains are low; in the second place that they would have to be phenomenally high to produce a change in temperature sufficient to condense the air about them and so cap them with cloud; and in the third place that the air is not dense enough to support clouds, anyway. Nevertheless a most singular phenomenon was seen by Mr. Douglass on November 24, a bright detached projection, for which from measurement he deduced a height of thirty miles. This would seem to have been cloud. With regard to its enormous height, it is not to be forgotten that a few years ago, on the earth, phenomenal dust-clouds were observed as high as one hundred miles.

Something more in the line of the explicable was a phenomenon observed in 1879 and in 1881 by Schiaparelli. From October, 1879, to January, 1880, he noticed certain bright patches which appeared to surround the north pole in a sort of crown, the pole itself being invisible. In 1881 he saw the same ramifications again, in apparently the same place. At this latter opposition the north pole was much better placed for observation, and he was able to mark a curious subsequent action in these spots; for as time went on they gradually contracted toward the pole, till finally they consolidated into the north polar patch, which up to that time had been absent. The polar patch proper did not thus appear till more than a month after the vernal equinox of the northern hemisphere.

Here, then, we have a very curious phenomenon, a phenomenon which seems to indicate that the seasonal wave of change acts as a unit across the planet's face; that instead of a more or less continuous deposit of moisture at the pole, such as occurs on earth, Martian atmospheric conditions oblige such deposit to creep gradually with the season up into polar latitudes, where it appears first as a crown of frost, and does not envelop the pole and become a polar cap till it has got higher. No sooner has this happened than the advance of following warmer isotherms causes it to begin to melt. One deduction from this thin air we must, however, be careful not to make: that because it is thin it is incapable of supporting intelligent life. That beings constituted physically as we are would find it a most uncomfortable habitat is pretty certain. But lungs are not wedded to logic, and there is nothing in the world or beyond it to prevent, so far as we know, a being with gills, for example, from being a most superior person. A fish doubtless imagines life out of water to be impossible; and similarly, to argue that life of an order as high as our own, or higher, is impossible, because of less air to breathe than that to which we are locally accustomed, is, as Flammarion happily expresses it, to argue, not as a philosopher, but as a fish.

To sum up, now, what we know about the atmosphere of Mars: we have proof positive that Mars has an atmosphere; we have reason to believe that this atmosphere is very thin,—thinner at least by half than the air upon the summit of the Himalayas,—that in constitution it does not differ greatly from our own, and that it is relatively heavily charged with water vapor.

In the next paper I shall take up the question of water upon the planet.

Percival Lowell.

II.

THE WATER PROBLEM.

AFTER air, water. If Mars be capable of supporting life, there must be water upon his surface; for to all forms of life water is as vital a matter as air. To all organisms water is absolutely essential. On the question of habitability, therefore, it becomes all-important to know whether there be water on Mars.

Any one looking through a telescope at the planet, early last summer, would at once have been struck by the fact that its surface was diversified by markings in three colors,—white, blue-green, and reddish-ochre; the white lying in a great oval at the top of the disc. The white oval was the south polar ice-cap. In this polar cap our water problem takes its rise.

On the 31st of May, 1894, the south polar cap stretched, practically one unbroken waste of ice, over about fifty degrees of latitude; that is, it covered nearly the whole frigid zone. Although due, in all probability, to successive depositions of frost rather than snow, the result, both in appearance and in behavior, makes striking counterpart to the antarctic ice-sheet of our own earth. Its visible contour was almost perfectly elliptical, showing it to be, in truth, nearly circular. That it was already in active process of melting was evident from its slowly lessening size. It was the most interesting feature on the disc, being peculiarly well placed for observation, owing to the tilt of the polar axis; for the Martian south pole was at the time bowed toward the earth at an angle of 24°, a southern inclination which has not been equaled since 1877. The dip

27

of the pole displayed the snow-cap to great advantage, and enabled the metamorphosis it underwent to be specially well seen.

Through June and July the snows were melting very fast, at the rate of hundreds of square miles a day. Such waning of them under the summer sun has been regularly observed to take place for the past two hundred years. At every Martian summer they shrink away to next to nothing, as systematically as the Martian summer comes on,—an action on their part highly indicative of their character. But another bit of behavior in their immediate surroundings is yet more significant, and in the case of the southern hemisphere has never, apparently, been noticed before.

Practically at the first observation made at Flagstaff in June, there showed, bordering the edge of the snow, a narrow, dark blue band or ribbon of color encircling the cap. The band varied in width at different places, being widest where the blue-green areas to the north of it were widest, and narrowest where they were narrowest. Its greatest breadth was about two hundred and twenty miles, its least about one hundred. In two places it expanded into great bays, the more prominent of them being just above the largest blue-green area on the disc. That the width of this antarctic girdle was proportionate to the width of the blue-green areas below it is a highly suggestive fact. Both bays were blue, the larger and more striking one especially so, appearing in good seeing of a beautiful cobalt blue, like some Martian grotto of Capri.

Both the band and its bays were contrasted with the blue-green areas contiguous to them, somewhat in tint, but yet more in tone. They were bluer, and distinctly darker. This hinted at a difference of constitution, which hint was emphasized by the action of the band; for as the snow shrank back toward the pole the blue belt followed it, keeping pace with its retreat. Instead of remaining in the place where it had first appeared, as it must have done had it been a permanent marking upon the surface of the planet, it withdrew steadily southward, so as always to border the melting snow.

28

At about the same time a rift made its appearance in the midst of the ice-cap. On June 9, when on the meridian early in the morning it looked like a huge cart-track coming down toward one through the snow. It proved to be three hundred and fifty miles wide, and debouched into the dark encircling band. A second narrower rift ran into it near the centre of the cap.

On the same morning, about half past five o'clock, starlike points suddenly shone out upon the snow, between the great bay and the first rift. After shining there for a few minutes they as suddenly vanished. It is evident what these were,—not fabled flash-lights of the Martians, but the glint of snow-slopes tilted at just the angle to reflect the sun toward the earth. On subsequent mornings others appeared, not so brilliant, the position of the planet with regard to the earth having slightly shifted in the mean time. There is something romantic in the thought of these far-off glistens from other-world antarctic snows, and in the sight there is corroboration of the snow's character.

As the Martian spring progressed, the rifts spread, until at last they cut the ice-cap in two. The smaller portion then proceeded to disappear, while the larger shrank correspondingly in size. The relative times of disappearance of different parts of the cap give us some information about the relative elevations and depressions of the south circumpolar regions. In consequence, I have been able to construct a contour map of these polar portions of the planet. There are advantages in thus conducting polar expeditions astronomically. One not only lives like a civilized being through it all, but he brings back something of the knowledge he went out to acquire.

There has even been vouchsafed the realization of that dream of explorers, an open polar sea; for as the first rift widened, Professor W. H. Pickering marked a large lake develop in the midst of it, in position almost over the pole. It seems cynical of fate thus to permit a Martian open polar sea to be seen before granting our earthly explorers a similar sight.

29

As the snows dwindled in size, the blue band about them shrank to correspond. By August it was a barely discernible thread drawn round the tiny white patch which was all that remained of the enormous snow-fields of some months before. Finally, on October 13, the snow entirely disappeared, and the spot where it and its girdle, long since grown too small for detection, had been became one yellow stretch.

That the blue was water at the edge of the melting snow seems unquestionable. That it was of the color of water, that it so persistently bordered the melting snow, and that it subsequently vanished are three facts mutually confirmatory of this deduction. Professor W. H. Pickering made the polariscope tell the same tale; for, on scrutinizing the great bay through it, he found the light coming from the bay to be polarized. Now, to polarize the light it reflects is a property, as we know, of a water surface.

From all this we may conclude that we have here a polar sea, a real body of water. There is, therefore, water on the surface of Mars. We also mark that this body of water is ephemeral. It exists while the snow-cap is melting, and then it somehow vanishes. What becomes of it, and whether there be other bodies of water on the planet, either permanent or temporary, we shall now go on to inquire.

While it existed in any size, the polar sea was bordered on the north, all the way round and during all the time it was visible, by blue-green areas. These blue-green areas were strewn with several more or less bright regions, while below them came the great reddish-ochre stretches of the disc. Now, the blue-green areas have generally been considered to be seas, just as the reddish-ochre regions have been held to be land. That the latter are land there is very little doubt; not only land, but nothing but land,—land very pure and simple; that is, deserts. For they behave just as deserts should behave, chiefly by not behaving at all; remaining, except for certain phenomena to be specified later, unchangeable.

With the so-called seas, however, the case is different. Several important facts conspire to throw grave doubt, and worse, upon their aquatic character. To begin with, they are of every grade of tint,—a very curious feature for seas to exhibit, unless they were everywhere but a few feet deep; which again is a most singular characteristic for seas that cover hundreds of thousands of square miles in extent,—seas, that is, as big as the Bay of Bengal. The Martian surface would have to be amazingly flat for this to be possible. We know it to be relatively flat, but to be as flat as all this would seem to pass the bounds of credible simplicity. Here also Professor W. H. Pickering's polariscope investigations come in with effect, for he found the light from the supposed seas to show no trace of polarization. Hence these were probably not water.

In parenthesis we may here take notice of the absence of a certain phenomenon whose presence, apparently, should follow upon water surfaces such as the so-called seas would offer us. Although its absence is not perhaps definitive as to their marine character, it is certainly curious, and worth nothing. If a planet were covered by a sheet of water, that water surface would, mirror-like, reflect the sun in one more or less definite spot. Looked at from a distance, this spot would, were it bright enough, be seen as a high light on the dark background of the ocean about it. It would seem to be a fixed star at a certain point on the disc, the surface features rotating under it. The necessary position is easily calculated, and this shows that parts of the so-called seas, especially at oppositions like the last one, pass under the point. There remains merely the question of sufficient brilliancy in the spot for visibility; but as in the case of Mars its brilliancy should be equal to that of a star of the third magnitude, it would seem brilliant enough to be seen. No such starlike effect in such position has ever been noticed coming from the blue-green regions. From this bit of negative evidence, to be taken for what it is worth, we return again to what there is of a positive sort.

Not only do different parts of the so-called seas contrast in tint with one another, but the same part of the same sea varies in tint at different times. Schiaparelli noticed that, at successive oppositions, the same sea showed different degrees of darkness, and he suggested that the change in tone was dependent in some way upon the Martian seasons.

Observations at Flagstaff have demonstrated this to be the case, for it has been possible to see the tints occur consecutively. In consequence, we know not only that changes take place on the surface of Mars other than in the polar cap, and very conspicuous ones too, but that these are due to the changing seasons of the planet's year. We will now see what they look like.

To the transubstantiation of changes of the sort it is a prime essential that the drawings from whose comparison the contrast appears should all have been made by the same person, at the same telescope, under as nearly as possible the same atmospheric conditions, since otherwise the personal equation of the observer, the impersonal inequalities of instruments, and the special atmosphere of the station play so large a part in the result as to mask that other factor in the case, any change in the planet itself. How easily this masking is accomplished appears from drawings made by different observers of the same Martian features at substantially the same moment. Several interesting specimens of such personal peculiarities may be seen by the curious in Flammarion's admirable thesaurus, La Planete Mars. In some of these likenesses of the planet it is pretty certain that Mars would never recognize himself.

To have drawings simply swear at one another across a page is, in the interests of deduction, objectionable. For their testimony to be worth having they must agree to differ. If, therefore, Mars is to be many, his draughtsman must be one. So much, at least, is fulfilled by the drawings in which the changes now to be described are recorded; for they were all made by me, at the same instrument, under the same general atmospheric conditions. As the same personality enters all of them, it stands, as between them, eliminated from all, to increased certainty of deduction. Since, furthermore, the drawings were all made in the months preceding and following one opposition, change due to secular variation is reduced to a minimum. As a matter of fact, the changes are such as to betray their own seasonal character. They constitute a kinematical as opposed to a statical study of the planet's surface.

The changes are much more evident than might be supposed. Indeed, they are quite unmistakable. As for their importance, it need only be said that that deduction from them furnishes, in the first place, strong inference that Mars is a living world, subject to an annual cycle of surface growth, activity, and decay; and shows, in the second place, that this Martian yearly round of life must differ in certain interesting particulars from that which forms our terrestrial experience. The phenomena evidently make part of a definite chain of changes of annual development. So consecutive, and, in their broad characteristics, apparently so regular are these changes that I have been able to find corroboration of what appears to be their general scheme in drawings made at previous oppositions. In consequence, I believe it will be possible in future to foretell, with something approaching the certainty of our esteemed weather bureau's prognostications, not indeed what the weather will be on Mars,—for, as we have seen, it is more than doubtful whether Mars has what we call weather to prognosticate,—but the aspect of any part of the planet at any given time.

The changes in appearance now to be chronicled refer, not to the melting of the polar snows, except as such melting forms the necessary preliminary to what follows, but to the subsequent changes in look of the surface itself. To their exposition, however, the polar phenomena become inseparable adjuncts, since they are inevitable ancillaries to the result.

With the familiar melting of the snow-cap begins the yearly round of the planet's life. With the melting of our own arctic or antarctic cap might similarly be said to begin the earth's annual activity. But here at the very outset there appears to be one important difference between the two planets. On the earth the relation of the melting of the polar snows to the awakening of surface activity is a case of *post hoc* simply; on Mars it seems to be a case of *propter hoc* as well. For, unlike the earth, which has water to spare, and to which, therefore, the unlocking of its polar snows is a matter of no direct economic value, Mars is apparently in straits for the article, and has to draw on its polar reservoir for its annual supply. Upon the melting of its polar cap, and the transference of the water thus annually set free to go its rounds, seem to depend all the seasonal phenomena on the surface of the planet.

The observations upon which this deduction is based extend over a period of nearly six months, from the last day of May to the 22d of November. They cover the regions from the south pole to about latitude forty north. That changes analogous to those recorded, differing, however, in details, occur six Martian months later in the planet's northern hemisphere is proved by what Schiaparelli has seen; for though the general system is, curiously, one for the whole planet, the particular character of different parts of the surface alters the action there to some extent.

For an appreciation of the meaning of the changes, it is to be borne in mind throughout that the vernal equinox of Mars' southern hemisphere occurred on April 17, 1894; the summer solstice of the same hemisphere on August 31; and its autumnal equinox on February 7, 1895.

On the 31st of May, therefore, it was toward the end of April on Mars. The south polar cap was, as we have seen, very large, and the polar sea in proportion. That the polar sea was the darkest and the bluest marking on the disc implies that it was, at all events, the deepest body of water on the planet, whether the other so-called seas were seas or not. This polar sea plays *deus ex machina* to all that follows.

So soon as the melting of the snow was well under way, long straits, of deeper tint than their surroundings, made their appearance in the midst of the dark areas. I did not see them come, but as I afterward saw them go, it is evident that they must have come. They were already there on the last day of May. The most conspicuous of them lay between Noachis and Hellas, in the Mare Australe. It began in the great polar bay, and thence traversed the Mare Erythraeum to the Hour-Glass Sea (Syrtis Major). The next most conspicuous one started in the other bay, and came down between Hellas and Ausonia. Although these straits were distinguishably darker than the seas through which they passed, the seas themselves were then at their darkest. The fact that these straits traversed the seas suffices to raise a second doubt as to the genuineness of seas; the first suspicion as to their true character coming from their being a little off color,—not so blue, that is, as what we practically know to be water, the polar sea, although even that must be anything but deep. It will appear later that in all probability the straits too are impostors, and that we see is in neither case water.

The appearance of things at this initial stage of the Martian Nile-like inundation last June was most destructive to modern maps of Mars, for all the markings between the south polar cap and the continental coast-line seemed with one consent to have as nearly as might be obliterated themselves.

It was impossible to fix any definite boundaries to the south temperate chain of islands, so indistinguishably did the light areas and the dark ones merge into each other. What was still more striking, the curious peninsulas which connect the continent with the chain of islands to the south of it, and form so singular a feature of the planet's geography, were invisible. One continuous belt of blue-green stretched from the Syrtis Major to the Columns of Hercules.

For some time the dark areas continued largely unchanged in appearance; during, that is, the earlier and most extensive melting of the snow-cap. After this their history became one long chronicle of fading out. Their lighter parts grew lighter, and their darker ones less dark. For, to start with, they were made up of many tints; various shades of blue-green interspersed with hints of orange-yellow. The gulfs and bays bordering the continental coast were the darkest of these markings; the long straits between the polar sea and the Syrtis Major were the next deepest in tone.

The first marked sign of change was the reappearance of Hesperia. Whereas in June it had been practically non-existent, by August it had become perfectly visible and in the place where it is usually depicted. In connection with its reappearance two points are to be noted: first, the amount of the change, for Hesperia is a stretch of land over two hundred miles broad by six hundred miles long; and, secondly, the fact that its previous invisibility was not due to any sort of obscuration. The persistent clear-cut character of the neighboring coast-line during the whole transformation showed that nothing in the way of mist or cloud had at any time hidden the peninsula from view. A something was actually there in August which had not been there in June.

As yet nothing could be seen of Atlantis. It was not until the 30th of October that I caught sight of it. About the same time, the straits between the islands, Xanthus, Scamander, Ascanius, and Simois, came out saliently dark, a darkness due to contrast. The line of south temperate islands and their separate identity were then for the first time apparent.

Meanwhile, the history of Hesperia continued to be instructive. From having been absent in June and conspicuous in August, it returned in October to a mid-position of visibility. Vacillating as these fluctuations in appearance may seem at first sight, they were really quite consistent; for they were due to one progressive change in the same direction, a change that was manifested first in Hesperia itself, and then in the regions round about it. From June to August, Hesperia changed from a previous blue-green, indistinguishable from its surroundings, to yellow, the parts adjacent remaining much as before. As a consequence, the peninsula stood out in marked contrast to the still deep blue-green regions by its side. Later, the surroundings themselves faded, and their change had the effect of once more partially obliterating Hesperia.

While Hesperia was thus causing itself to be noticed, the rest of the south temperate zone, as we may call it for identification's sake, was unobtrusively pursuing the same course. Whereas in June all that part of the disc comprising the two Thyle, Argyre II., and like latitudes was chiefly blue-green, by October it had become chiefly yellow. Still further south, what had been first snow, and then water, turned to ochre land.

Certain smaller details of the change that came over the face of the dark regions at the time were as curious as they were marked. For example, the Fastigium Aryn, the tip of the triangular cape which, by jutting out from the continent, forms the forked bay called the Sabaeus Sinus, and which, because of its easy identification, has been selected for the zero meridian of Martian longitudes, began in October to undergo strange metamorphosis. On October 15 it shot out a sort of tail southward. On the 16th this tail could be followed all the way to Deucalionis Regio, to which it made a bridge across from the continent, thus cutting the Sabaeus Sinus completely in two.

Another curious causeway of the same sort made its appearance in November, connecting the promontory known as Hammonis Cornu with Hellas. Both of these necks of orange-ochre were of more or less uniform breadth throughout.

The long, dark streaks that in June had joined the Syrtis Major to the polar sea had nearly disappeared by October; in their southern parts they had vanished completely, and they had very much faded in their northern ones. The same process of fading uncovered certain curious rhomboidal bright areas in the midst of the Syrtis Major.

It will be seen that the extent of these changes was enormous. Their size, indeed, was only second in importance to their character; for it will also have been noticed that the changes were all in one direction. A wholesale transformation of the blue-green regions into orange-ochre ones was in progress upon that other world.

What can explain so general and so consecutive a change in hue? Water suggests itself; for a vast transference of water from the pole to the equator might account for it. But there are facts connected with the change which seem irreconcilable with the idea of water. In the first place, Professor W. H. Pickering found that the light from the great blue-green areas showed no trace of polarization. This tended to strengthen a theory put forth by him some years ago, that the greater part of the blue-green areas are not water, but something which at such a distance would also look blue-green, namely, vegetation. Observations at Flagstaff not only confirm this, but limit the water areas still further; in fact, practically do away with them entirely. Not only do the above polariscopic tests tend to this conclusion, but so does the following observation of mine in October.

Toward the end of October, a strange, and, for observational purposes, a distressing phenomenon took place. What remained of the more southern dark regions showed a desire to vanish, so completely did those regions proceed to fade in tint throughout. This was first noticeable in the Cimmerian Sea, then in the Sea of the Sirens, and in November in the Mare Erythraeum about the Lake of the Sun. The fading steadily progressed until it had advanced so far that in poor seeing the markings were almost imperceptible, and the planet presented a nearly uniform ochre disc.

This was not a case of obscuration; for in the first place it was general, and in the second place the coast-lines were not obliterated. The change, therefore, was not due to clouds or mist.

What was suggestive about the occurrence was that it was unaccompanied by a corresponding increase of blue-green elsewhere. It was not simply that portions of the planet's surface changed tint, but that, taking the disc in its entirety, the whole amount of the blue-green upon it had diminished, and that of the orange-yellow had proportionally increased. Mars looked more Martian than he had in June. The canals, indeed, began at the same time to darken; but, highly important as this was for other reasons, the whole area of their fine lines and associated patches did not begin to make up for what the dark regions lost.

If the blue-green color was due to water, where had all the water gone? Nowhere on the visible parts of the planet; that is certain. Nor could it very well have gone to those north circumpolar regions hid from view by the tilt of the disc; for there was no sign of a growing north polar cap, and, furthermore, Schiaparelli's observations upon that cap show that there should not have been. As we saw in the last paper, he found that it developed late, apparently one month or so after the vernal equinox of its hemisphere, whereas at the time the above change occurred it was not long after that hemisphere's winter solstice.

But if, instead of being due to water, the blue-green tint had been due to leaves and grasses, just such a fading out as was observed should have taken place as autumn came on, and that without proportionate increase of green elsewhere; for the great continental areas, being desert, are incapable of supporting vegetation, and therefore of turning green.

There is thus reason to believe that the blue-green regions of Mars are not water, but, generally at least, areas of vegetation; from which it follows that Mars is very badly off for water, and that the planet is dependent on the melting of its polar snows for practically its whole supply.

Such scarcity of water on Mars is just what theory would lead us to expect. Mars is a smaller planet than the earth, and therefore is relatively more advanced in his evolutionary career. He is older in age, if not in years; for whether his birth as a separate world antedated ours or not, his smaller size, by causing him to cool more quickly, would necessarily age him faster. But as a planet grows old, its oceans, in all probability, dry up, the water retreating through cracks and caverns into its interior. Water thus disappears from its surface, to say nothing of what is being continually imprisoned by chemical combination. Signs of having thus parted with its oceans we see in the case of the moon, whose so-called seas were probably seas in their day, but have now become old sea-bottoms. On Mars the same process is going on, but would seem not yet to have progressed so far, the seas there being midway in their career from real seas to arid depressed deserts; no longer water surfaces, they are still the lowest portions of the planet, and therefore stand to receive what scant water may yet travel over the surface. They thus become fertilized, while higher regions escape the freshet, and remain permanently barren. That they were once seas we have something more than general inference to warrant us in believing.

There is a certain peculiarity about the surface markings of Mars, which is pretty sure to strike any thoughtful observer who examines the planet with a two or a three inch object-glass,—their singular sameness night after night. With quite disheartening regularity, each evening presents him with the same appearance he noted the evening before,—a dark band obliquely belting the disc, strangely keeping its place in spite of the nightly progression of the meridians ten degrees to the east, in consequence of our faster rotation gaining on the slower rotation of Mars. By attention, he will notice, however, that the belt creeps slowly upwards towards the pole in all other respects. Then suddenly some night he finds that it has slipped bodily down, to begin again its Sisyphus-like, inconclusive spiral climb.

Often as this rhomb-line must have been noticed, no explanation of it has ever, to my knowledge, been given. Yet so singular an arrangement points to something other than chance. Suspicion of its non-fortuitous character is strengthened when it is scanned through a bigger glass. Increase of aperture discloses details that help explain its significance. With sufficient telescopic power, the continuity of the dark belt is seen to be broken by a series of parallel peninsulas or semi-peninsulas that jut out from the lower edge of the belt, all running with one accord in a southeasterly direction, and dividing the belt into a similar series of parallel dark areas. Such oblong areas are the Mare Tyrrhenum, the Mare Cimmerium, the Mare Sirenum, and those unnamed straits that stretch southeasterly from the Aurorae Sinus, and Margaritifer Sinus, and the Sabaeus Sinus. The islands and peninsulas trending in the same direction are Ausonia, Hesperia, Cimmeria, Atlantis, Pyrrhae Regio, Deucalionis Regio, and the two causeways from the Fastigium Aryn and Hammonis Cornu. It will further be noticed that these areas lie more nearly north and south as they lie nearer the pole, and curve in general to the west as they approach the equator.

With this fact noted, let us return to the water formed by the melting of the ice-cap, at the time it is produced around the south pole. We may be sure it would not stay there long. No sooner liberated from its winter fetters than it would begin, under the pull of gravity, to run toward the equator. The reason why it would flow away from the pole is that it would find itself in unstable equilibrium where it was. Successive depositions of frost would have piled up a mound of ice which, so long as it remained solid, cohesion would keep in that unnatural position, but the moment it changed to a liquid this would flow out on all sides, seeking its level. Once started, its own withdrawal would cause the centre of gravity to shift away from the pole, and this would pull the particles of the water yet more toward the equator. Each particle would start due north; but its course would not continue in that direction, for at each mile it traveled it would find itself in a lower latitude, where, owing to the rotation of the planet, the surface would be whirling faster toward the east, inasmuch as a point on the equator has to get over much more space in twenty-four hours than one nearer the pole. In short, supposing there was no friction, the surface would be constantly slipping away from under the particle toward the east. As a result, the northerly motion of the particle would be continually changing with regard to the surface into a more and more westerly one. If the surface were not frictionless, friction would somewhat reduce the westerly component, but could never wholly destroy it without stopping this particle.

We see, therefore, that any body, whether solid, liquid, or gaseous, must, in traveling away from the pole of a sphere or spheroid, necessarily deviate to the west as it goes on, if the spheroid itself revolve, as Mars does, in the opposite direction.

This inevitable trend induced in anything flowing from the pole to the equator is precisely the one that we notice stereotyped so conspicuously in the Martian south temperate markings. Here, then, we have at once a suspiciously suggestive hint that they once held water, and that that water flowed.

Corroborating this deduction is the fact that the northern sides of all the dark areas are very perceptibly darker than the southern ones; for the northern side is the one which a descending current would plough out, since it is the northern coasts that would be constantly opposing the current's northerly inertia. Consequently, although at present the descending stream be quite inadequate to such task, it still finds its way, from preference, to these lowest levels, and makes them greener than the rest.

Though seas no longer, we perceive, then, that there is some reason to believe the so-called seas of Mars to have been seas in their day, and to be at the moment midway in evolution from the seas of the earth to the seas of the moon.

Now, if a planet were at any stage of its career able to support life, it is probable that a diminishing water supply would be the beginning of the end of that life, for the air would outlast the available water. Those of its inhabitants who had succeeded in surviving would find themselves at last face to face with the relentlessness of fate,—a scarcity of water constantly growing greater, till at last they would all die of thirst, either directly or indirectly; for either they themselves would not have water enough to drink, or the plants or animals which constituted their diet would perish for lack of it,—an alternative of small choice to them, unless they were conventionally particular as to their mode of death. Before this lamentable conclusion was reached, however, there would come a time in the course of the planet's history when water was not yet wanting, but simply scarce and requiring to be husbanded; when, for the inhabitants, the one supreme problem of existence would be the water problem,—how to get water enough to sustain life, and how best to utilize every drop of water they could get.

Mars is, apparently, in this distressing plight at the present moment, the signs being that its water supply is now exceedingly low. If, therefore, the planet possess inhabitants, there is but one course open to them in order to support life. Irrigation, and upon as vast a scale as possible, must be the all-engrossing Martian pursuit. So much is directly deducible from what we have learned recently about the physical condition of the planet, quite apart from any question as to possible inhabitants. What the physical phenomena assert is this: if there be inhabitants, then irrigation must be the chief material concern of their lives.

At this point in our inquiry, when direct deduction from the general physical phenomena observable on the planet's surface shows that were there inhabitants there a system of irrigation would be an all-essential of their existence, the telescope presents us with perhaps the most startling discovery of modern times,—the so-called canals of Mars. These strange phenomena, together with the inferences to be drawn from them, will form the subject of the next paper.

III.

CANALS.

IN the last paper we saw how badly off for water Mars, to all appearance, is; so badly off that any inhabitants of that other world would have to irrigate to live. As to the actual presence there of such folk, the broad physical characteristics of the planet have nothing to say beyond a general expression of acquiescence, but they do have something very vital to say about the conditions under which alone their life could be led. They show that in these Martian minds there would be one question paramount to all the local labor, women's suffrage, and Eastern questions put together,—the water question. How to procure water enough to support life would be the great communal problem of the day.

If Mars were the earth, we might well despair of detecting signs of any Martians for some time yet. Across the gulf of space that separates us from Mars, an area thirty miles wide would just be perceptible as a dot. It would, in such case, be hopeless to look for evidence of folk. Anything like London or New York, or even Chicago in anticipation, would be too small to be seen. So sorry a figure does man cut upon the earth he thinks to own. From the standpoint of forty millions of miles' distance, probably the only sign of his presence here would be such semi-artificialities as the great grain-fields of the West when their geometric patches turned with the changing seasons from ochre to green, and then from green to gold. By his crops we should know him,—a telltale fact of importance because probably the more so on Mars.

For Mars is not the earth. Conditions hold there which would necessitate a different state of things, inorganic and organic, apparently a much more artificial one. If cultivation there be, it must be cultivation upon a much more systematic scale, due in large part to a system of irrigation; just as any Martians must be quite different physically from men.

Now, at this point in our investigation, when the broad features of Mars disclose conditions which imply irrigation as their organic corollary, we are suddenly confronted on the planet's face with phenomena so startlingly suggestive of this very thing as to seem the uncanny realization of the deduction. Indeed, so amazingly lifelike is their appearance that, had we possessed our present knowledge of the planet's physical condition before, we might almost have predicted what we see as criterion of the presence of living beings. What confronts us is this:—

When the great continental areas, the reddish-ochre portions of the disc, are attentively examined in sufficiently steady air, their desert-like ground is seen to be traversed by a network of fine, straight dark lines. The lines start from points on the coast of the blue-green regions, commonly well-marked bays, and proceed direct to other equally well-marked points in the middle of the continent. At these latter termini the lines meet, very surprisingly, other lines that have come there from different starting-points in a similarly definite manner. And this state of things exists all over the reddish-ochre regions.

All the lines, with the exception of a few that are curved in a regular manner, are absolutely straight from one end to the other. They are arcs of great circles, taking the shortest distance between their termini. The lines are as fine as they are straight. As a rule, they are of scarcely any perceptible breadth, seeming on the average to be less than a Martian degree, or between twenty and thirty miles, wide. Some are broader; some even finer, possibly not above fifteen miles across. Their length, not their breadth, renders them visible; for though at such a distance we could not distinguish a dot less than about thirty miles in diameter, we could see

45

a line of much less breadth, because of its length. Speaking generally, however, the lines are all of comparable width.

Still greater uniformity is observable in the different parts of the same line; for each line maintains its individual width throughout. Although at and near the point where it leaves the dark regions, or the Solis Lacus,— for the same phenomenon appears there,—some slight enlargement seems to take place, after it has fairly started on its course it remains substantially of the same size from one end to the other. As to whether the lines are even on their edges or not, I should not like to say, but the better they are seen, the more even they look. It is not possible to affirm positively on the point, as they are practically nearer one dimension than two.

On the other hand, their length is usually great, and in some cases enormous. A thousand or fifteen hundred miles may be considered about the average. The Ganges, for example, which is not a long one as Martian canals go, is about 1450 miles in length. The Brontes, one of the newly discovered, radiating from the Gulf of the Titans, extends over 2400 miles. Among really long ones, the Eumenides, with its continuation the Orcus, the two being in truth one line, runs 3540 miles from the point where it leaves the Phoenix Lake to the point where it enters the Trivium Charontis; throughout this great distance, nearly equal to a diameter of the planet, deviating neither to the right nor to the left from the great circle upon which it set out. On the other hand, the shortest line is the Nectar, which is only about 250 miles in length; sweetness being, according to Schiaparelli its christener, as short-lived on Mars as elsewhere.

That with very few exceptions the lines all follow arcs of great circles is proved: first, by the fact that when near the centre of the disc they show as straight lines; second, that when seen toward its edges they appear curved, in keeping with the curvature of a spherical surface viewed obliquely; third, that when the several parts of some of the longer lines are plotted upon a globe they turn out to lie in one great circle. Apparent straightness throughout is only possible in short lines. For a very long arc upon the surface of a revolving globe tilted toward the observer to appear straight in its entirety it must lie in certain positions. It so chances that these conditions are fulfilled by the canal called the Titan. The Titan starts from the Gulf of the Titans, in south latitude 20°, and runs due north almost exactly upon the 169th meridian for an immense distance. I have followed it over 2300 miles down the disc to about 43° north, as far as the tilt of the planet's axis would permit. As the rotation of the planet swings it round, it passes the central meridian of the disc simultaneously throughout its length, and at that moment comes out strikingly straight, a substantialized meridian itself.

Although each line is the arc of a great circle, the direction taken by this great circle may be any whatsoever. The Titan, as we have seen, runs nearly due north and south. Certain canals crossing this run, on the contrary, almost due east and west. There are others, again, belting the disc at well-nigh every angle between the two. Nor is there any preponderance, apparently, for one direction as against any other. This indifference to direction is important as showing that the rotation of the planet has no direct effect upon the inclination of the canals.

But, singular as each line looks to be by itself, it is the systematic network of the whole that is most amazing. Each line not only goes with wonderful directness from one point to another, but at this latter spot it contrives to meet, exactly, another line which has come with like directness from quite a different starting-point. Nor do two only manage thus to rendezvous. Three, four, five, and even seven will similarly fall in on the same spot,—a sociability which, to a greater or less extent, takes place all over the surface of the planet. The disc is simply a network of such intersections. Sometimes a canal goes only from one intersection to another; more commonly it starts with right of continuation, and, after reaching the first rendezvous, goes on in unchanged course to several more.

The result is that the whole of the great reddish-ochre portions of the planet is cut up into a series of spherical triangles of all possible sizes and shapes. What their number may be lies quite beyond the possibility of count at present; for the better our own air, the more of them are visible. About four times as many as are down on Schiaparelli's chart of the same regions have been seen at Flagstaff. But before proceeding further with a description of these Martian phenomena, the history of their discovery deserves to be sketched, since it is as strange as the canals themselves.

The first hint the world had of their existence was when Schiaparelli saw some of the lines in 1877, now eighteen years ago. The world, however, was anything but prepared for the revelation, and, when he announced what he had seen, promptly proceeded to disbelieve him. Schiaparelli had the misfortune to be ahead of his times, and the yet greater misfortune to remain so; for not only did no one else see the lines at that opposition, but no one else succeeded in doing so at subsequent ones. For many years fate allowed Schiaparelli to have them all to himself, a confidence he amply repaid. While others doubted, he went from discovery to discovery. What he had seen in 1877 was not so very startling in view of what he afterward saw. His first observations might well have been of simple estuaries, long natural creeks running up into the continents, and so cutting them in two. His later observations were too peculiar to be explained even by so improbable a configuration of the Martian surface. In 1879, the *canali*, as he called them (channels, or canals, the word may be translated, and it is in the latter sense that he now regards them), showed straighter and narrower than they had in 1877: this not in consequence of any change in them, but from his own improved faculty of detection; for what the eye has once seen it can always see better a second time. As he gazed they appeared straighter, and he made out more. Lastly, toward the end of the year, he observed, one evening, what struck even him as a most startling phenomenon, the twinning of one of the canals: two parallel canals suddenly showed where but a single one had showed before. The paralleling was so perfect that he suspected optical illusion.

He could, however discover none by changing his telescopes or eyepieces. The phenomenon, apparently, was real.

At the next opposition he looked to see if by chance he should mark a repetition of this strange event, and went, as he tells us, from surprise to surprise; for one after the other of his canals proceeded startlingly to become two, until some twenty of them had thus doubled. This capped the climax to his own wonderment, and, it is needless to add, to other people's incredulity; for nobody else had yet succeeded in seeing the canals at all, let alone seeing them double. Undeterred by the general skepticism, he confirmed, at each fresh opposition, his previous discoveries; which, in view of the fact that no one else did, rather tended in astronomical circles to the opposite result.

For nine years he labored thus alone, having his visions all to himself. It was not till 1886 that any one but he saw the canals. In April of that year Perrotin at Nice first did so. The occasion was the setting-up of the great Nice glass of twenty-nine inches aperture. In spite of the great size of the glass, however, a first attempt resulted in nothing but failure. So did a second, and Perrotin was on the point of abandoning the search altogether when, on the 15th of the month, he suddenly detected one of the canals, the Phison. His assistant, M. Thollon, saw it immediately afterward. After this they managed to make out several others, some single, some double, substantially as Schiaparelli had drawn them; the slight discrepancies between their observations and his being, in point of fact, the best of confirmations.

Since then other observers have contrived to detect the canals, the list of the successful increasing at each opposition, although even now their number might almost be told on one's hands and feet. The fact that so few men have yet seen these lines is due to poor air. That in ordinary atmosphere the canals are not easy objects is certain; while for the detection of their peculiar fineness and straightness a steady air is essential. So also is attentive perception on the part of the observer, size of aperture being distinctly a secondary matter. That Schiaparelli discovered the canals with an 8 1/2 object-glass, and that the 26-inch at Washington has refused to show them to this day, are facts that speak with emphasis on the point.

Although skepticism as to the existence of the so-called canals seems now pretty well dispelled, disbelief still makes a desperate stand against their peculiar appearance, dubbing accounts of their straightness and duplication as sensational, whatever that may mean in such connection; for that they are both straight and double, as described, is certain,—a statement I make after having seen them instead of before doing so, as is the case with the gifted objectors. Doubt, however, will not wholly cease till more people have seen them, which will not happen till the importance of atmosphere in the study of planetary detail is more generally appreciated than it is to-day. To look for the canals with a large instrument in poor air is like trying to read a page of fine print kept dancing before one's eyes, and increase of magnification increases the motion. Advance in our study of other worlds depends upon choosing the very best atmospheric sites for our observatories.

As we shall now have to call these Martian things by their names,— our names, that is,—it may be well to consider cursorily the nomenclature which has been evolved on the subject. Unfortunately, the planet has been quite too much benamed,—benamed, indeed, out of all recognition. There are no less than five or six systems current for its general topographical features. The result is that it has become something of a specialty just to know the names. The Syrtis Major, for example, appears under the following aliases: the Syrtis Major, the Mer du Sablier, the Kaiser Sea, the Northern Sea, to say nothing of translations of these, such as the Hour-Glass Sea. After which ample baptism it is a trifle disconcerting to have the sea turn out, apparently, not to be a sea at all. Everybody has tried his hand at naming the planet, first and last; naming a thing being man's nearest approach to creating it. Proctor made a chart of the planet, and named it thoroughly; Flammarion drew another chart, and also named it thoroughly, but differently; Green made a third map, and gave it a third

50

set of names; Schiaparelli followed with a fourth, and furnished it with a brand-new set of his own; and finally W. H. Pickering found it necessary to give a few new names, just for particularization. To know, therefore, what part of the planet anybody means when he mentions it, one has to keep in his head enough names for five worlds. To cap which, it is to be remarked that not one of them is the thing's real—that is, its Martian—name, after all!

Fortunately, with the canals matters are not so desperate, because so few people have seen them. Schiaparelli's monopoly of the sight pleasingly prevented, in their case, christening competition. What is more, he named them very judiciously and most picturesquely after mythologic river names. Where he got his names is another matter. Whether he started by being as learned in such lore as he afterward became may well be doubted. Certainly one of the greatest discoveries made at Flagstaff has been the discovery of the meaning of Schiaparelli's names; some of them still defying the penetrating power of the ordinary encyclopaedia. Among them are classical mythologic ones of the class known only to that himself mythical character, Macaulay's every schoolboy, which speaks conclusively for their reconditeness. Others, I firmly believe, even that omniscient schoolboy can never have heard of. Want of space here precludes instances; but as a simple example I may say that the translation to Mars of the Phison and the Gehon, the two lost rivers of Mesopotamia, satisfactorily accounts for their not being found on earth by modern explorers.

With due mental reservation as to their meaning, I have adopted Schiaparelli's names, and where it has been necessary to name newly discovered canals have conformed as closely as possible to his general scheme. If even in an instance or two I have hit upon names that are incomprehensible, I shall feel that I have not disgraced my illustrious predecessor. For a brand-new thing no name is so good as one whose meaning nobody knows, except one that has no meaning at all.

Schiaparelli's scheme embraces all the other Martian features as well as the canals, and the same poetic imagination pervades the whole. For example, the central promontory of what used to be known as Dawes' Forked Bay, a prominent point, since it has for some time been used as the zero meridian for Martian longitudes, he calls the Fastigium Aryn. The Fastigium Aryn was, it appears, the cupola of the world, a mythic spot supposed to be the absolute centre of the earth regarded as a plane in mid-heaven,—a point midway between the north and south, the east and west, the zenith and nadir; an eminently suitable name, indeed, for the origin of longitudes and the beginning of time.

To return now to the objects of so much human incredulity. The first point worth noting about them is that their actual existence is quite beyond question; the second, that the better they are seen, the odder they look. Observations at Arequipa in 1892 not only confirmed Schiaparelli's, but extended the canal system considerably both in quantity and in character; observations last year at Flagstaff extended it still further, so that now we know of about half as many more canals as are down on Schiaparelli's chart, and of certain phenomena connected with them no less peculiar, to say the least, than themselves. What these strange dependencies are we will note after we have considered the canals.

So far we have regarded the canals only statically, so to speak; that is, we have sketched them as they would appear to any one who observed them in sufficiently steady air, once, and once only. But this is far from all that a systematic study of the lines will disclose. Before, however, entering upon this second phase of their description, we may pause to note how, even statically regarded, the aspect of the lines is enough to put to rest all the theories of purely natural causation that have so far been advanced to account for them. This negation is to be found in the supernaturally regular appearance of the system, upon three distinct counts: first, the straightness of the lines; second, their individually uniform width; and third, their systematic radiation from special points.

On the first two counts we observe that the lines exceed in regularity any purely natural regularity of which we commonly have cognizance. Physical processes never, so far as we know, produce perfectly regular results; that is, results in which irregularity is not also plainly discernible. Disagreement amid conformity is the inevitable outcome of the many factors simultaneously at work. From the orbits of the heavenly bodies to philotaxis and human features, this diversity in uniformity is apparent. As a rule, the divergences, though small, are quite perceptible; that is, the lack of absolute uniformity is comparable to the uniformity itself, and not of the negligible second order of unimportance. In fact, it is by the very presence of uniformity and precision that we suspect things of artificiality. It was the mathematical shape of the Ohio mounds that suggested mound-builders; and so with the thousand objects of every-day life. Too great regularity is in itself the most suspicious of circumstances that some finite intelligence has been at work.

If it be asked how, in the case of a body so far off as Mars, we can assert sufficient precision to imply artificiality, the answer is two fold: first, that the better we see these lines, the more regular they look; and second, that the eye is quicker to perceive irregularity than we commonly note. It is indeed surprising to find what small irregularities will shock the eye.

The third count is, if possible, yet more conclusive. That the lines form a system; that, instead of running any-whither, they join certain points to certain others, making thus, not a simple network, but one whose meshes connect centres directly with one another, is striking at first sight, and loses none of its peculiarity on second thought. For the intrinsic improbability of such a state of things arising from purely natural causes becomes evident on consideration.

Were lines drawn haphazard over the surface of a globe, the chances are ever so many to one against more than two lines crossing each other at any point. Simple crossings of two lines would of course be common in something like factorial proportion to the number of lines, but that any other line should contrive to cross at the same point would be a coincidence whose improbability only a mathematician can properly appreciate, so very great is it. If the lines were true lines, without breadth, the chances against such a coincidence would be infinite; and even had the lines some breadth, the chances would be enormous against a rendezvous. In other words, we might search in vain for a single instance of such encounter. On the surface of Mars, however, instead of searching in vain, we find the thing occurring *passim*; this *a priori* most improbable rendezvousing proving the rule, not the exception. Of the crossings that are best seen, almost all are meeting-places for more than two canals.

To any one who had not seen the canals, it would at once occur that something of the same improbability might be fulfilled by cracks radiating from centres of explosion or fissure. But such a supposition is at once negatived by the uniform breadth of the lines, a uniformity impossible in cracks, whose very mode of production necessitates their being bigger at one end than the other. We see examples of what might result from such action in the cracks that radiate from Tycho, in the moon, or, as we now from Professor W. H. Pickering's observations, from the craterlets about it. These cracks bear no resemblance whatever to the lines on Mars. They look like cracks; the lines on Mars do not. Indeed, it is safe to say that the Martian lines would never so much as suggest cracks to any one. Lastly, the different radiations fit into one another absolutely, an utter impossibility were they radiating rifts from different centres.

In the same way, we may, while we are about it, show that the lines cannot be several other things which they have, more or less gratuitously, been taken to be. They cannot, for example, be rivers; for rivers could not be so obligingly of the same size at source and mouth, nor would they run from preference on arcs of great circles. To do so, practically invariably, would imply a devotion to pure mathematics not common in rivers. They may, in some few instances, be rectified rivers, which is quite another matter. Glaciation cracks are equally out of the question: first, for the causes above mentioned touching cracks in general; and second,

54

because there is, unfortunately, no ice where they occur. Nor can the lines be furrows ploughed by meteorites,—another ingenious suggestion,—since in order to plough, invariably, a furrow from one centre to another, without either swerving from the course or overshooting the mark, the visitant meteorite would have to be carefully trained to the business.

Such are the chief purely natural theories of the lines, excluding the idea of canals,—theories advanced by persons who have not seen them. No one who has seen the lines well has or could advance them, inasmuch as they are not only disproved by consideration of the character of the lines, but instantly confuted by the mere look of them.

Schiaparelli supposes the canals to be canals, but of geologic construction. He suggests, however, no explanation of how this is possible; so that the suggestion is not, properly speaking, a theory. That eminent astronomer further says of the idea that they are the work of intelligent beings, "Io mi quardero bene dal combattere questa supposizione la quale nulla include d'impossibile." (I should carefully refrain from combating this supposition, which involves no impossibility.) In truth, no natural theory has yet been advanced which will explain these lines, while recent observations furnish material that seems to render artificial construction probable.

After so much necessary digression upon what the canals are not, we will resume our inquiry as to what they are.

So far we have considered their aspect at any one time, and we have seen that it is such as to defy natural explanation, and to hint that in these lines we are regarding something other than the outcome of purely natural causes. Indeed, such is the first impression upon getting a good view of them. How instant this inference is becomes patent from the way in which drawings of the canals are received by incredulously disposed persons. The straightness of the lines is unhesitatingly attributed to the draughtsman. Now it is to be remembered that accusation of design, if it prove inapplicable to the draughtsman, devolves *ipso facto* upon the canals.

We come next to a consideration of their successive appearances night after night, and month after month. After the fundamental fact that such curious phenomena as the canals are visible is the scarcely less important one that they are not always so. At times the canals are invisible, and this invisibility is real, not apparent; that is, it is not an invisibility due to distance or obscuration of any kind between us and them, but an actual invisibility due to the condition of the canal itself. With our present optical means, at certain seasons they cease to exist. For aught we can see, they simply are not there.

That distance is not responsible for the disappearance of the canals is shown by their relative conspicuousness at different times. It is not always when Mars is nearest to us that the canals are best seen. On the contrary, they show a sublime disregard for mere proximity. This is evidenced both by the changes in appearance of any one canal and by the changes in relative conspicuousness of different canals. Some instances of the metamorphosis will reveal this conclusively. For example, during the end of August and the beginning of September, at this last opposition, the canals about the Lake of the Sun were conspicuous, while the canals to the north of them were almost invisible. In November the relative intensities of the two sets had distinctly changed: the southern canals were much as before, but the northern ones had most perceptibly darkened.

Another instance of the same thing was shown in the case of the canals to the north of the Sinus Titanum when compared with those about the Solis Lacus. In August the former were but faintly visible; in November they had become evident; and yet, during this interval, little change in conspicuousness had taken place in the canals in the Solis Lacus region.

With like disregard of the effect due to distance, the canals to the east of the Ganges showed better at the November presentation[1] of that

1   A presentation of any part of the planet is the occasion when that part of the
      disc is turned toward the observer. Many causes combine to make the face
      presented each night vary, but the chief one is that the earth rotates about
      forty-one minutes faster than Mars, and consequently gains a little less than

region than they had at the October one, although the planet was actually farther off at the later date, in the proportion of 21 to 18.

A more striking instance of the irrelevancy of distance in the matter was observed in the same region by Schiaparelli in 1877. It is additionally interesting as practically dating his discovery of the canals. In early October of that year, on the evenings of the 2d and the 4th, he tells us, under excellent definition, and with the diameter of the planet's disc 21" of arc, the continental region between the Pearl-Bearing Gulf and the Bay of the Dawn was quite uniformly, nakedly bright, and destitute of suspicion of markings of any sort. A like state of things was the case with the same region at its next presentation, on the 7th of November. Four months later, when the diameter of the disc had been reduced by distance to 5".7, or, in other words, when the planet had receded to four times its previous distance from the earth, the canal called the Indus appeared, perfectly visible, in the region mentioned. At the next opposition, in 1881, similar effects occurred; the canals in this region remaining obstinately invisible while the planet was near the earth, and then coming out conspicuously when it had gone farther away. Distance, therefore, is not, with the canals, the great obliterator.

As to their veiling by Martian cloud or mist, there is no evidence of any such obscuration. The coast line of the dark areas appears as clear-cut when the canals are invisible as when they become conspicuous.

A canal, then, alters its visibility for some reason connected with itself. It grows into recognition from intrinsic cause. But during all its metamorphoses, in one thing, and in one thing only, it remains fixed,—in position. Temporary in appearance, the canals are apparently permanent in place. Not only do they not change in position during one opposition; they seem not to do so from one opposition to another. The canals I

---

ten degrees on him daily. After about thirty-seven days, therefore, the two planets again present the same face to each other at the same hour. Their first appearance is a matter of the Martian time of year.

have observed this year agree quite within the errors of observation with those figured on Schiaparelli's chart. In general they conform to their representations, and failure to do so is explicable not only by errors of observation, but by certain other facts. First, by seasonal variation in the canals themselves; the visibility or invisibility of a canal combined with the visibility or invisibility of a neighbor being capable of producing strange permutations in the region observed.

The Araxes is a case in point. On Schiaparelli's chart there is but one original Araxes and one great and only Phasis. But it turns out that these do not possess the land all to themselves. No less than five canals traversing the region, including the Phasis itself, were visible this year at Flagstaff, and I have no doubt there are plenty of others waiting to be discovered. These cross one another at all sorts of angles. Unconscious combination of them is quite competent to give a turn to the Araxes one way or the other, and make it curved or straight at pleasure.

Unchangeable, apparently, in position, the canals are otherwise among the most changeable features of the Martian disc. From being invisible, they emerge gradually, for some reason inherent in themselves, into conspicuousness. In short, phenomenally at least, they grow. The order of their coming carries with it a presumption of cause, for it synchronizes with the change in the Martian seasons.

To start with, the visible development of the canal system follows the melting of the polar snows. Not until such melting has progressed pretty far do any of the canals, it would seem, become perceptible.

Secondly, when they do appear, it is, in the case of the southern hemisphere, the most southern ones that become visible first. Last June, when the canals were first seen, those about the Lake of the Sun and the Phoenix Lake were easier to make out than any of the others. Now, this region is the part of the reddish-ochre continent, as we may call it, that lies nearest the south pole. It extends into the blue-green regions as far south as 40 ° of south latitude. Nor do any so-called islands—that is, smaller reddish-ochre areas—stand between it and the pole. It lies first exposed, therefore, to any water descending toward the equator from the melting of the polar cap.

Having once become visible, these canals remained so, becoming more and more conspicuous as the season advanced. By August they had darkened very perceptibly. As yet those in other parts of the planet were scarcely more visible than they had been two months before. Gradually, however, others became evident, farther and farther north, till by October all the canals bordering the north coast of the dark regions were recognizable; after which the latter, in their turn, proceeded to darken,—a state of things which continued up to the close of my observations toward the end of November.

The order in which the canals came out hinted that two factors were operative to the result, latitude and proximity to the dark regions. Other things equal, the most southern ones showed first; beginning with the Solis Lacus region, and continuing with those about the Sea of the Sirens and the Titan Gulf, and so northward down the disc. Other things were not, however, always equal in the way of topographical position. Notably was this the case with the areas to the west of the Syrtis Major, which developed canals earlier than their latitudes would warrant. Now, to the Syrtis Major descend from the pole the great straits spoken of before, which, although not in their entirety water, are probably lands fertilized by a thread of water running through them. They connect the polar sea with the Syrtis Major in a tolerably straight line.

The direction of the canal also affects its time of appearance, though to a less extent. Canals running north and south, such as the Gorgon, the Titan, the Brontes, and the like, became visible, as a rule, before those running east and west. Especially was this noticeable in the more northern portions of the disc. Time of appearance was evidently a question of latitude tempered by ease of communication.

After the canals had appeared, their relative intensities changed with time, and the change followed the same order in which the initial change from invisibility to visibility had taken place. A like metamorphosis happened to each in turn from south to north, in accordance with, and continuance of, the seasonal change that affected all the blue-green areas.

To account for these phenomena, the explanation that at once suggests itself is, that a direct transference of water takes place over the face of the planet, and that the canals are so many waterways. This explanation has the difficulty of involving enormously wide canals. There is another objection to it: the time taken would appear to be too long, for some months elapsed between the apparent departure of the water from the pole and its apparent advent in the equatorial regions; furthermore, each canal did not darken all at once, but gradually. We must therefore seek some explanation which accounts for this delay. Now, when we do so, we find that the explanation advanced above for the blue-green areas explains also the canals, namely, that what we see in both is, not water, but vegetation; for if the darkening be due to vegetation, time must elapse between the advent of the water and its perceptible effects,—time sufficient for the flora to sprout. If, therefore, we suppose what we call a canal to be, not the canal proper, but the vegetation along its banks, the observed phenomena stand accounted for. This suggestion was first made some years ago by Professor W. H. Pickering.

That what we see is not the canal proper, but the line of land it irrigates, disposes incidentally of the difficulty of conceiving a canal several miles wide. On the other hand, a narrow, fertilized strip of country is what we should expect to find; for, as we have seen, the general physical condition of the planet leads us to the conception, not of canals constructed for waterways,—like our Suez Canal,—but of canals dug for irrigation purposes. We cannot, of course, be sure that such is their character, appearances being often highly deceitful; we can only say that, so far, the supposition best explains what we see. Further details of their development point to this same conclusion.

In emerging from invisibility into evidence, the canals first make themselves suspected rather than seen, as broad, faint streaks smooching the disc. Such effect, however, seems to be an optical illusion, due to poor air and the difficulty inherent in detecting fine detail; for on improvement in the seeing I have observed these broad streaks contract to fine lines, not sensibly different in width from what they eventually become.

The parts of the canals which are nearest the dark areas show first, the line extending sometimes for a few hundred miles into the continent, sometimes for a thousand or more; then, in course of time, the canal becomes evident in its entirety. Complete visibility takes place soon after the canal has once begun to show, although it show but faint throughout.

This tendency to being seen *in toto* is more strikingly displayed after a canal has attained its development. It is then not commonly seen in part. Either it is not seen at all, owing to the seeing not being good enough, or it is visible throughout its length from one junction to another.

Apart from their extension, the growth of the canals consists chiefly in depth of tint. They darken rather than broaden,—a fact which tends to corroborate their vegetal character; for that long tracts of country should be thus simultaneously flooded all over to a gradually deepening extent is highly unlikely, while a growth of vegetation would deepen in appearance in precisely the way that the darkening takes place.

As for color, the lines would seem to be of the same tint as the blue-green areas. But, owing to their narrowness, this is only an inference. I have never chanced to see them of distinctive color.

At this point it is probable that a certain obstacle to such wholesale construction of canals, however, will arise in the mind of the reader, namely, the thought of mountains; for mountains are by nature antagonistic to canals. Only the Czar of all the Russias—if we are to credit the account of the building of the Moscow railway—would be capable of running a canal regardless of topography. Nor will the doings at our own antipodes help us to conceive such construction; for though the Japanese irrigate hillsides, the water in the case comes from slopes higher yet, whereas on Mars it does not.

Indeed, for the lines to contain canals we must suppose either that mountains prove no obstacles to Martians, or else that there are practically no mountains on Mars. For the system seems sublimely superior to possible obstructions in the way; the lines running, apparently, not where they may, but where they choose. The Eumenides-Orcus, for example, pursues the even tenor of its unswerving course for nearly 3500 miles. Now, it might be possible so to select one's country that one canal should be able to do this; but that every canal should be straight, and many of them fairly comparable in length, seems to be beyond the possibility of contrivance.

In this dilemma between mountains on the one hand and canals on the other, a certain class of observations most opportunely comes to our aid; for, from observations which have nothing to do with the lines, it turns out that the surface of the planet is, in truth, most surprisingly flat. How this is known will most easily be understood from a word or two upon the manner in which astronomers have learnt the heights of the mountains in the moon.

The heights of the lunar mountains are found from measuring the lengths of the shadows they cast. As the moon makes her circuit of the earth, a varying amount of her illuminated surface is presented to our view. From a slender sickle she grows to a full moon, and then diminishes again to a crescent. The illuminated portion is bounded by a semi-circle on the outside, and by a semi-ellipse on the inner. The semicircle is called her limb, the semi-ellipse her terminator. The former is the edge we see because we can see no further; the latter, the line upon her surface where the sun is just rising or setting. Now, as we know, the shadows cast at sunrise or sunset are very long, much longer than the objects that cast them are high. This is due to the obliquity at which the light strikes them; the same effect being produced by any sufficiently oblique light, such as an electric light at a distance. Imperceptible in themselves, the heights become perceptible by their shadows. A road illuminated by a distant arc light gives us a startling instance of this; the smooth surface taking on from its shadows the look of a ploughed field.

It is this indirect kind of magnification that enables astronomers to measure the lunar mountains, and even renders such vicariously visible to the naked eye. Every one has noticed how ragged and irregular the inner edge of the moon looks, while her outer edge seems perfectly smooth. In one place it will appear to project beyond the perfect ellipse, in another to recede from it. The first effect is due to mountain tops catching the sun's rays before the plains about them; the other, to mountain tops further advanced into the lunar day, whose shadows still shroud the valleys at their feet. Yet the elevations and depressions thus rendered so noticeable vanish in profile on the limb.

Much as we see the moon with the naked eye do we see Mars with the telescope. Mars being outside of us with regard to the sun, we never see him less than half illumined, but we do see him with a disc that lacks of being round,—about what the moon shows us when two days off from full. It is when he is in quadrature—that is, a quarter way round the celestial circle from the sun—that he shows thus, and we see him then with the telescope at closer range than we ever see the moon. When we so observe him, we notice at once that his terminator, or inner edge, presents a very different appearance from the lunar one. Instead of looking like a saw, it looks comparatively smooth, like a knife. From this we know that, relatively to his size, he has no elevations or depressions upon his surface comparable to the lunar peaks and craters.

His terminator, however, is not absolutely perfect. Irregularities are to be detected in it, although much less pronounced than those of the moon. His irregularities are of two kinds. The first, and by all odds the commonest phenomenon consists in showing himself on occasions surprisingly flat; not in this case an inferable flatness, but a perfectly apparent one. In other words, his terminator does not show as a semi-ellipse, but as an irregular polygon. It looks as if in places the rind had been pared off. The peel thus taken from him, so to speak, is from twenty to forty degrees wide, according to the particular part of his surface that shows upon the terminator at the time.

Now it is a significant fact that this paring of his disc appears usually where the dark regions are coming into view or passing out of sight, according as it is the sunrise or the sunset terminator that is presented to observation. And even in the few cases where it is not coincident with them, it is never far removed from their position. Two causes undoubtedly combine to produce the effect. One of them is irradiation. It is a well-known fact that bright bodies look larger than they are, probably because of the sympathetic vibration of the rods in the retina adjoining those directly affected. A familiar instance of the effect is the seemingly wizened look of the old moon seen in the new moon's arms. The lusty young moon seems a sixth the broader of the two. The same thing would appear in the case of the Martian terminator; a bright area would seem to project beyond a dark one. This accounts for a part of the loss. The other part is doubtless due to an actual depression in the Martian surface. Thus from the appearance of the terminator comes corroboration of the lower level at which we found reason (in the last paper) to suppose the dark markings upon the planet to lie.

That these long parings do not always coincide with the dark areas may help confirm, paradoxical as it sounds, their real depression; for it is only the relative, not the actual height that is projected on the terminator, and a more elevated area, if sloping at the proper angle, would be projected as a depression beside a lower one, in spite of being the higher surface of the two. It may also, however, not be due to this cause, but to the presence of an actually elevated district; verdure, such as a forest, standing on high land.

Such long, low depressions are characteristic of the Martian terminator, which is thus in kind quite unlike the lunar one. In addition to them there are elevations, some long and low, some short and sharp. Both are relatively rare. Of the former variety Professor W. H. Pickering discovered two striking specimens. Each looked to be, and probably was, a plateau, very level on top, and sloping more or less equally on both sides. Of the short and sharp variety Mr. Douglass has detected some noteworthy instances; but whether they mean high dust cloud or mountains is not yet predicable. Mr. Douglass has very systematically observed the Martian terminator at every longitude, and is now busy upon a contour map of the planet. His map may enable us to say something more definite as to whether the canals traverse low regions from preference or not. But certain it is that Mars is a flat world; devoid, as we may note incidentally, of summer resorts, since it possesses, apparently, neither seas nor hills. To canals we will now return.

The canals so far described all lie in the bright reddish-ochre portions of the disc,—those parts which bear every appearance of being desert. But Mr. Douglass has made the discovery that they are not the only part of the planet thus privileged. He finds, in the very midst of the dark regions themselves, straight, dark streaks not unlike in look to the canals, and still more resembling them in the systematic manner in which they run. For they reproduce the same rectilinear arrangement that is so striking a characteristic of their bright-area fellows. He has succeeded, indeed, in thus triangulating all the more important dark areas. What is more, he finds that these canals in the dark regions end at the very points at which the others begin, so that they make continuations of them.

This fact is another telltale circumstance as to the true character of the so-called seas; for that the seas should be traversed by permanent dark lines is incompatible with a fluid constitution. But the lines are even more suggestive from a positive than they are from a negative standpoint. That they make continuations of the lines in the bright regions shows that the two sets are causally connected, and affords strong presumption that this causal relation is the very one demanded by the theory of irrigation. For if the canals in the bright regions be strips of vegetation irrigated by a canal (too narrow to be itself visible at our distance), and there be a scarcity of water upon the surface of the planet, the necessary water would have to be conducted to the mouths of the canals across the more permanent areas of vegetation, thus causing bands of denser verdure athwart them, which we should see as dark lines upon the less dark background.

Before passing on to certain other phenomena connected with the canals of like significance, we may note here an *obiter* dictum of the irrigation theory of some slight corroborative worth; for if a theory be correct, it will not only fit all the facts, but at times go out of its way to answer questions. Such the present one seems to do. If the seas be seas, and the canals canals, we stand confronted by the problem how to make fresh-water canals flow out of salt-water seas. General considerations warrant us in believing that the Martian seas, like our own, would contain salts in solution, while irrigation ditches, there as here, should flow fresh water to be most effective, and we seem committed to the erection of distilleries upon a gigantic scale. But if, on the contrary, the seas be not seas, but areas of vegetation, the difficulty vanishes at once; for if the planet be dependent upon the melting of its polar snows for its spring freshet, the water thus produced must necessarily be fresh, and the canals be directly provided with the water they want. The polar sea is a temporary body of water, formed anew each year, not a permanent ocean; consequently there is no chance for saline matter to collect in it. From it, therefore, fresh water flows, and, like our rivers, gathers nothing to speak of in the way of salt before it is drawn off into the canals.

We now come to some phenomena connected with the canals, of the utmost suggestiveness. I have said that the junctions held in a twofold way the key to the unlocking of the mystery of the canals; in the first place, in the fact that such junctions exist. The second and more important reason remains to be given, for it consists in what we find at those junctions. These phenomena will form the subject of the next paper.

IV.

OASES.

SUGGESTIVE of irrigation as the strange network of lines that covers the surface of Mars appears to be, the suggestion takes on more definite shape yet with the last addition to our knowledge of the planet's surface detail,—the recognition of a singularly correlated system of spots.

The canals, as we have seen, show a remarkable attachment to their kind. Not content with such casual meetings as chance would afford them in the course of their long careers, they make a point of rendezvousing as often and in as great numbers as possible. Indeed, the ingenuity with which they manage to combine unswerving rectitude with meetings by the way grows more and more marvelous, the more one studies it. The meeting-places, or junctions, evidently possess an attraction for the canals. The crossings, in fact, seem to be the end and aim of the whole system; the canals, but means to that end. So much is at once inferable from the great intrinsic improbability that such crossings can be due to chance.

The inference receives, apparently, striking verification from a something which turns out to exist at these junctions. This something shows itself as a round or ovate spot. To such spot, planted there in the midst of the desert, do the neighboring canals converge.

Dotted all over the reddish-ochre ground of the great desert stretches of the planet, the so-called continents of Mars, are an innumerable number of dark circular or ovate spots. They appear, furthermore, always in intimate association with the canals. They constitute so many hubs to which the canals make spokes. These spots, together with the canals that lead to them, are the only markings to be seen anywhere on the continental regions. Otherwise the great reddish-ochre areas are absolutely bare; of that pale fire-opal hue which marks our own deserts seen from far.

That these two things, straight lines and roundish spots, should, with our present telescopic means, be the sole markings to appear on the vast desert regions of the planet is suggestive in itself.

Another significant fact as to the character of either marking is the manifest association of the two. In spite of the great number of the spots, not one of them stands isolate. There is not a single instance of a spot that is not connected by a canal to the rest of the dark areas. This remarkable inability to stand alone shows that the spots and the canals are not unrelated phenomena, for were there no tie between them they must occasionally exist apart.

Nor is this all. There is, apparently, no spot that is not joined to the rest of the system, not only by a canal, but by more than one; for though some spots, such as the Fountain of Youth, have appeared at first to be provided with but a single canal connection, later observation has revealed concurrence in the case. The spots are, therefore, not only part and parcel of the canal system, but terminal phenomena of the same.

They are, generally speaking, more difficult features to see than the canals. In consequence, they have been among the most recent details to be made out upon the planet's surface. It was not until 1892, at Arequipa, that they were seen in anything like their real numbers. Of them, indeed, are the forty lakes found by Professor W. H. Pickering. This year, at Flagstaff, still others have been discovered, to detection of their character, as I think.

In the first place, as I have said, there appears to be no spot that has not two or more canals running to it; in the second place, I find, reversely, that apparently no canal junction is without its spot. Such association is a most tell-tale circumstance. I believe the rule to have no exception. The more prominent junctions all show spots; and with regard to the less conspicuous ones, it is to be remembered that, as the canals are more easy to make out than the spots, the relative invisibility of the latter is to be expected. From which it would seem that the spots are fundamental features of the junctions, and that for a junction to be spotless is, from its very nature, an impossibility.

Next to their regularity of position is to be remarked their regularity of form. Their typical shape seems to be circular; for the better the atmosphere, the rounder they look. Under poor seeing they show as irregular patches smooching the disc, much as the canals themselves show as streaks; the spots differing from the canals in being thicker and not so long. As the seeing improves, the patches differentiate themselves into round dots and connecting lines. Such is the shape of the spots associated with single canals; that is, canals not double. In the case of the double canals, the spots look like rectangles with the corners rounded off. One of the most striking of all of them is the Trivium Charontis, which is nearly square.

Now it will be noticed that these shapes are as unnatural as they are definite, and that they all agree in one peculiarity: they are all convex, not concave, to the entering canals. They are not, therefore, mere enlargements of the canals, due to natural causes; for were the spots enlargements of the canals at their crossing points they should be more or less star-shaped, or concave to the canals, whereas they are round, or roundish rectangles,—that is convex to the same. Such convexity negatives, at the outset, their being purely natural outgrowths of the canals.

The majority of the spots are from 120 to 150 miles in diameter; thus presenting a certain uniformity in size as well as in shape. There are some smaller ones, not more than 75 miles across, or less.

To the spot category belong all the markings other than canals to be seen anywhere on the continental deserts of the planet, from the great Lake of the Sun, which is 540 miles long by 300 miles broad, to the tiny Fountain of Youth, which is barely distinguishable as a dot. That all are fundamentally of a kind is hinted at by their shape and emphasized by their character, a point to which we shall come in a moment.

To this end, we will start with an account of where and how they begin to show; for, like the canals, they are not permanent markings, but temporary phenomena. It is in the region about the Solis Lacus that they appear first. The Solis Lacus, or Lake of the Sun, is perhaps the most striking marking on Mars. It is an oval spot in lat. 28° S., with its greater diameter nearly perpendicular to the meridians, and encircled by an elliptical ring of reddish-ochre land, which in turn is bordered on the south by the blue-green regions of the south temperate zone. The whole configuration is such as to simulate a gigantic eye which uncannily turns round upon one as the planet slowly revolves. It is so conspicuous a feature of the disc that it has been recognized for a great many years. The resemblance to an eye is further borne out by a cordon of canals that surround it on the north. Upon this cordon, composed chiefly of the Araxes and the Agathodaemon, are beaded a number of spots, two of them, the Phoenix and the Tithonus lakes, being conspicuously prominent. Closer scrutiny reveals several more of the same sort, only smaller. These are all interconnected by a network of canals. Now just as it is in this region that the canals first show, so likewise is it here that the spots first make their appearance.

Although it was here that at this last opposition the spots were first seen, it was not here that their character and purpose became apparent. It was not until later in the season, when the Eumenides-Orcus began to give evidence of being yet more peculiarly beaded, that the true nature of the spots suggested itself.

The Eumenides-Orcus is a very long and important canal, connecting the Phoenix Lake with the Trivium Charontis. It is so long, 3540 miles from one end of it to the other, that although it starts in lat. 16° N., and ends in lat. 12° S., it belts the disc not many degrees inclined to the equator. For a great distance it runs parallel to the northern coast of the Sea of the Sirens. From this coast several canals strike down to it;

some stopping at it, others continuing on down the disc. Especially is the western end of the sea, called the Gulf of the Titans, a point of departure for canals; no less than six of them, and doubtless more, leaving the gulf in variously radiating directions. At the place where these canals severally cross the Eumenides-Orcus, I began in November to see spots. I also saw others along the Pyriphlegethon, an important canal leading in a more northerly direction from the Phoenix Lake; along the Gigas, a great canal running from the Gulf of the Titans all the way to the Lake of the Moon; and along other canals in the same region. I then noticed that the spots to the north of the Solis Lacus region had darkened, since August, relatively to the more southern ones. In short, I became aware both of a great increase in the number of spots, and of an increase in tint in the spots previously seen.

It was apparent that the spots were part and parcel of the canal system, and that in the matter of varying visibility they took after the canals,— chronologically, very closely after them; for a comparison of the two leads me to believe that the spots make their appearance subsequent, although but little subsequent, to the canals which conduct to them.

Furthermore, the spots, like the canals, grow in conspicuousness with time. Now when we consider that nothing, practically, has changed between us and them in the interval; that there has been no symptom of cloud or other obscuration, before or after, over the place where they eventually appear, we are led to the conclusion that, like the canals, they grow.

Indeed, in the history of their development the two features seem quite similar. Both grow, and both follow the same order and method in their growth. Both are affected by one progressive change that sweeps over the face of the planet from the pole to the equator, and then from the equator toward the other pole. In the case of the southern hemisphere, it is, as we have just seen, the most southern spots, like the most southern canals, that appear first after the melting of the polar snows. Then gradually others begin to show farther and farther north. The quickening of the spots, like the quickening of the canals, is a seasonal affair. But there is more in it than this. It takes place in a manner to imply that something more immediate than the change in the seasons is concerned in it; immediate not in time, but in relation to the result. A comparison of the behavior of three spots—the Phoenix Lake, the spot at the junction of the Iris and the Gigas, at the upper extremity of Ceraunius, and a spot where the Steropes, a newly found canal, and the Nilus meet—will serve to point out what this something is. The Phoenix Lake lies in lat. 17° S., the upper Ceraunius in lat 12° N., and the spot on the Steropes in lat. 28° N. In August of last year, the first of these markings was very conspicuous, the second but moderately so, while the third was barely discernible. By November, the Phoenix Lake had become less salient, Ceraunius relatively more so, and the spot on the Steropes nearly as evident as Ceraunius had formerly been. In the Martian calendar, the August observation corresponded to our 20th of June, the November one to our 1st of August, of the southern hemisphere; or to our 20th of December and 1st of February, respectively, of the northern one. All three spots were practically within the equatorial regions. Now, on earth, no such marked progression in seasonal change occurs within the tropics. With us, it is to all intents and purposes equally green there the year through. On Mars it is not. Clearly, some more definite factor than the seasons enters into the matter upon our neighbor world.

That this factor is water seems, from the behavior of the blue-green areas generally, to be pretty certain. But just as the so-called seas are undoubtedly not seas, nor the canals waterways, so the spots are not lakes. Their mode of growth, so far as it may be discerned, confirms this conclusion. Apparently, it is not so much by an increase in size as by a deepening in tint that they gradually become recognizable. They start, it would seem, as big as they are to be, but faint in tone, premonitory shades of their future selves. They then proceed to substantialize by darkening in tint throughout. Now, to deepen thus in color with one consent all over would be a peculiar thing for a lake to do. For had the lake appreciable depth to start with, it should always be visible; and had it not, its bed would have to be phenomenally level to permit of its being all flooded at once. If, however, the spots be not bodies of water, but areas of verdure, their deepening in tint throughout is perfectly explicable, since the darkening would be the natural result of a simultaneous growth of vegetation. This inference is further borne out by the fact that to the spot class belong unquestionably those larger oval markings of which the Lake of the Sun is the most conspicuous example. For both are associated in precisely the same manner with the canal system. Each spot is a centre of canal connections in exactly the way in which the Solis Lacus or the Phoenix Lake itself is. But the light coming from the Solis Lacus and the Phoenix Lake showed, in Professor W. H. Pickering's observations, no sign of polarization such as a sheet of water should show, and such as the polar sea actually did show.

When we put all these phenomena together,—the presence of the spots at the junctions of the canals, their strangely systematic shapes, their seasonal darkening, and last, but not least, the resemblance of the great continental regions of Mars to the deserts of the earth,—a solution of their character suggests itself at once: to wit, that they are oases in the midst of that desert, and not wholly innocent of design; for, in number, position, shape, and behavior, the oases turn out as typical and peculiar a feature of Mars as the canals themselves.

Each phenomenon is highly suggestive considered alone, but each acquires still greater significance from its association with the other; for here in the oases we have an end and object for the existence of canals, and the most natural one in the world, namely, that the canals are constructed for the express purpose of fertilizing the oases. Thus the mysterious rendezvousing of the canals at these special points is at once explicable. The canals rendezvous so entirely in defiance of the doctrine of chances because they were constructed to that end. They are not purely natural developments, but cases of assisted nature, just as they look to be at first sight. This, at least, is the only explanation that fully accounts for the facts. Of course all such evidence of design may be purely fortuitous, with about as much probability, as it has happily been put, as that a chance collection of numbers should take the form of the multiplication table.

In addition to this general dovetailing of detail to one conclusion is to be noticed the strangely economic character of both the canals and the oases in the matter of form. That the lines should follow arcs of great circles, whatever their direction, is as unnatural from a natural standpoint as it would be natural from an artificial one; for the arc of a great circle is the shortest distance from one point upon the surface of a sphere to another. It would, therefore, if topographically possible, be the course to take to conduct water, with the least expenditure of time or trouble, from the one to the other.

The circular shape of the oases is as directly economic as is the straightness of the canals; for the circle is the figure which incloses the maximum area for the minimum average distance from its centre to any point situated within it. In consequence, if a certain amount of country were to be irrigated, intelligence would suggest the circular form in preference to all others, in order thus to cover the greatest space with the least labor. In the case of the double canals, the same labor-saving intent would lead as instantly to a rounded rectangle.

Even more markedly unnatural is another phenomenon of this most phenomenal system, of which almost every one has heard, and which almost nobody has seen,—the double canals.

To see them, however, all that is needed is a sufficiently steady air, a sufficiently attentive observer, and the suitable season of the Martian year. When these conditions are observed, the sight may be seen without difficulty, and is every whit as strange as Schiaparelli, who first saw it, has described it.

So far as the observer is concerned, what occurs is this: Upon a part of the disc where up to that time a single canal has been visible, of a sudden some night, in place of the single canal appear twin canals,—as like, indeed, as twins, if not more so, similar both in character and in inclination, running side by side the whole length of the original canal, usually for upwards of a thousand miles, of the same size throughout, and absolutely parallel to each other. The pair may best be likened to the twin rails of a railroad track. The regularity of the thing is startling.

In good air the phenomenon is quite unmistakable. The two lines are as distinct and as distinctly parallel as possible. No draughtsman could draw them better. They are thoroughly Martian in their mathematical precision. At the very first glance, they convey, like all the other details of the canal system, the appearance of artificiality. It may be well to state this here definitely, for the benefit of such as, without having seen the canals, indulge in criticism about them. No one who has seen the canals well—and the well is all-important for bringing out the characteristics that give the stamp of artificiality, the straightness and fineness of the lines—would ever have any doubt as to their seeming artificial, however he might choose to blind himself to the consequences. An element akin to the comic enters criticism based not upon what the critics have seen, but upon what they have not. Books are reviewed without being read, to prevent prejudice; but it is rash to carry the same admirable broad-mindedness into scientific subjects.

In detail the doubles vary, chiefly, it would seem, in the distance the twin lines lie apart. In the widest I have seen, the Ganges, six degrees separate the two; in the narrowest, the Phison, four degrees and a quarter,—not a very great difference between the extremes. Four degrees and a quarter on Mars amount to 156 miles; six degrees, to 220. These, then, are the distances between the centres of the twin canals. Each canal seems a little less than a degree wide, or about 30 miles in the narrower instances; in the broader, a little more than a degree, or about 45 miles. Between the two lines, in the cases where the gemination, as it is called, is complete, lies reddish-ochre ground similar to the rest of the surface of the bright regions. Deducting the two half-widths of the bordering canals, we have, therefore, from 120 to 175 miles of clear country between the paralleling lines.

The gemination of a canal is a phenomenon individual to the particular canal. Each canal differs from its neighbor not only in the distance the lines lie apart, but in the time at which the duplication occurs. The event seems to depend both upon general seasonal laws governing all the duplications, and upon causes intrinsic to the canal itself. Within limits, each canal doubles at its own good time and after its own fashion. For example, although it seems to be a rule that north and south canals double before east and west ones, nevertheless, of two north and south lines, one will double, the other will not, synchronously with a doubling running east and west; the same is true of those running at any other inclination.

Now this shows that the duplication is not an optical illusion at this end of the line; for, by any double refraction here, all the lines running in the same direction over the disc should be similarly affected, which they are not. On the contrary, there will be, say, two cases of doubling in quite different directions co-existent with several single canals.

Nor is there any probability of its being a case of double refraction at the other end of the line,—that is, in the atmosphere of Mars; for in that case it is hard to see why all the lines should not be affected, to say nothing of the fact that, to render such double refraction possible, we must call upon a noumenon to help us out, as we know of no substance capable of the quality upon so huge a scale. Furthermore, what is cogent to the observer, though of no particular weight with his hearers, the phenomenon has no look of double refraction. It looks to be, what it undoubtedly is, a double existence.

Strengthening this conclusion is the mode of development of the doubling. This appears to take place in two ways, although it is possible that the two are but different instances of one and the same process. Of the first kind, during this last opposition, the Ganges was an example.

The Ganges was in an interesting protoplasmic condition during the whole of last summer. About to multiply by fission, it was not at first evident how this would take place. Hints of gemination were visible when I first looked at it in August. It showed then as a very broad but not dark swath of dusky color, of nearly uniform width from one extremity to the other, with sides suggestively even throughout. It is probable that they were then, as afterward, parallel, and that the slight convergence apparent at the bottom was due simply to foreshortening. The swath ran thus N. N. W. all the way from the Gulf of the Dawn to the Lake of the Moon. By moments of better seeing its two sides showed darker than its middle; that is, it was already double in embryo, with a dusky middle-ground between the twin lines.

In October the doubling had sensibly progressed. The double visions were more frequent, and the ground between the twin lines had grown lighter. By November the doubling was unmistakable, and the mid-clarification had become nearly complete. It is to be remarked that the doubling did not involve the Fons Juventae and the canal leading to it, both of which lay well to the right of the Ganges. The space included between the East and West Ganges was very wide, some six degrees. The canals themselves were, so far as could be seen, quite similar, and about a degree, or 37 miles wide. Both started in the Gulf of the Dawn, and ran down to the lower Lake of the Moon, one entering each side of the lake, or oasis. Two thirds of the way down both similarly touched the sides of another oasis, an upper Lacus Lunae. The whole length of each was 1200 miles.

Except for fleeting suspicions of gemination, and for possible doublings like the parallelism of the Styx and the Hades, the next canal to show double was the Nectar, which was so seen by Mr. Douglass on October 4, and under still better seeing, a few minutes later, the doubling was detected by him extending straight across the Solis Lacus. In the Solis Lacus this was evidently a case of mid-clarification. What occurred in the Nectar seems more allied to the second class of manifestations, such as happened later with the Euphrates and the Phison.

Glimpses of a dual state in these canals were caught during the summer and autumn, but it was not till the November presentation of the region that they came unmistakably twinned. On the 18th of that month, just as the twilight was fading away, the air being very still and the definition exceptional, so soon as the sunset tremors subsided, the Euphrates and its neighbor the Phison showed beautifully doubled, exactly like two great railroad tracks with bright ground between, each set extending down the disc for a distance of 1600 miles.

After that evening, whenever the seeing was good enough, they continued to present the same appearance. Now with them no process of midway clarification, such as had taken place in the Ganges, had previously made itself manifest. They had indeed not been very well defined before duplication occurred, but apparently sufficiently so not to hide such broadening had it taken place; for though the twin canals were not as far apart as the two Ganges, they were quite comparably distant, being, instead of six, about four and a quarter degrees from each other. Evidently, the process was, in the case of the Euphrates at least, under way in October, and even earlier, but was not well seen because the twin canals were not yet dark enough.

There seem, I may remark parenthetically, to be two other double canals in the region between the Syrtis Major and the Sabaeus Sinus, one to the east of the Phison, and another between the Phison and the Euphrates, both debouching at the same points as the Phison and the Euphrates themselves.

On the 19th of November I suspected duplication in the Typhon, another canal in the same region. It looked to be double, with dusky ground between.

On the 21st I similarly suspected the Jamuna and the Nilokeras. Both looked broad and dusky, with very ill-defined condensation at the sides.

But the seeing was not superlative. On the 22d I brought my observations to an end, in consequence of having to return East.

Exactly what takes place, therefore, in this curious process of doubling I cannot pretend to say. It has been suggested that a progressive ripening of vegetation from the centre to the edges might cause a broad swath of green to become seemingly two. There are facts, however, that do not tally with this view. For example, the Ganges was always broad, but fainter, not narrower, earlier in the season. The Phison, on the other hand, went through no such process. Indeed, we are here very much in the dark, certainly very far off from what does take place in Martian canal gemination. Perhaps we may learn considerably more about it at the next opposition. At this the tendril end of our knowledge of our neighbor we cannot expect hard wood.

To return now from these outposts of investigation to our main subject matter. We have seen what shows at one end of the canals, their inner end, namely, the oases. But it seems that there is also something exceptional at the other. At the mouth of each canal, at the edge of the so-called seas, appears a curious dark spot of the form of a half-filled angle; the sort of a mark with which one checks items on a list. Its form is singularly appropriate, according to mundane ideas, for it appears before the canal itself is visible, as if to mark the spot where the canal will eventually be. It lies in the so-called seas, and looks to be of the same color as they, but deeper in tint.

All the canals that debouch into the dark regions are provided with these terminal triangles, except those that lead out of long estuaries, like the Nilosyrtis, the Hiddekel, the Gehon, and so forth. The double canals are provided with twin triangles. That the triangular patches are phenomena connected with the canals is evident from the fact that they never appear elsewhere. What exact purpose they serve is not so clear, but it would seem to be that of reservoirs or relay stations for the water before it enters the canals; what we see, upon this supposition, being not the station or reservoir itself, but the specially fertile area round it.

That, in addition to being in a way oases themselves, they serve some such purpose as the above is further hinted at by two facts: first, that whereas the oases develop, apparently, after the canals leading to them, the triangular spots develop before the canals that lead out of them; second, Mr. Douglass finds that it is in them that the canals in the dark regions terminate. They are the end of the one system at the same time that they are the beginning of the other. They would, therefore, seem to be waystations of some sort on the road taken by the water from the polar cap to the equator.

Paralleling in appearance the oases in the bright regions are round spots that occur at the junctions of the canals in the dark ones. Speaking figuratively, these are the heads of the nails in the coffin of the idea that the seas are seas; since, if the blue-green color came from water, there could not be permanent darker dots upon it connected by equally dark streaks. Speaking unfiguratively, this shows that the whole system of canals and specially fertilized spots is not confined to the deserts, but extends in a modified form over the areas of more or less vegetation.

One of these specially fertile spots, situated upon the borderland betwixt the dark and the light regions, has a picturesque history. It lies at the head of the Margaritifer Sinus, or Pearl-Bearing Gulf, so named some years ago by Schiaparelli; the name having been given by him to the gulf quite fortuitously. But it turns out that the gulf was prophetically named, for there in it is this round spot which makes terminus to a short canal connecting it with the lower end of the western Sabaeus Sinus, and probably also terminus to a long canal coming from the Chrysorrhoas, across both branches of the Ganges. Diving into the depths of space has thus brought up the pearl from the bottom of the gulf.

We thus perceive that the blue-green areas are subjected to the same engineering system as the bright ones. In short, no part of the planet is allowed to escape from the all-pervasive trigonometric spirit. If this be Nature's doing, she certainly runs her mathematics into the ground.

To review, now, the chain of reasoning by which we have been led to regard it probable that upon the surface of Mars we see the effects of local intelligence: we find, in the first place, that the broad physical conditions of the planet are not antagonistic to some form of life; secondly, that there is an apparent dearth of water upon the planet's surface, and therefore, if beings of sufficient intelligence inhabited it, they would have to resort to irrigation to support life; thirdly, that there turns out to be a network of markings covering the disc precisely counterparting what a system of irrigation would look like; and, lastly, that there is a set of spots placed where we should expect to find the lands thus artificially fertilized, and behaving as such constructed oases should. All this, of course, may be a set of coincidences, signifying nothing; but the probability seems the other way. As to details of explanation, any we may adopt will undoubtedly be found, on closer acquaintance, to vary from the actual Martian state of things; for any Martian life must differ markedly from our own.

The fundamental fact in the matter is the dearth of water. If we keep this in mind, we shall see that many of the objections that spontaneously arise answer themselves. The supposed Herculean task of constructing such canals disappears at once; for if the canals be dug for irrigation purposes, it is evident that what we see and call, by ellipsis, the canal is not really the canal at all, but the strip of fertilized land bordering it,—the thread of water in the midst of it, the canal itself, being far too small to be perceptible. In the case of an irrigation canal seen at a distance, it is always the strip of verdure, not the canal, that is visible, as we see in looking from afar upon irrigated country on the earth.

Startling as the outcome of these observations may appear at first, in truth there is nothing startling about it whatever. Such possibility has been quite on the cards ever since the existence of Mars itself was recognized by the Chaldean shepherds, or whoever the still more primeval astronomers may have been. Its strangeness is a purely subjective phenomenon, arising from the instinctive reluctance of man to admit the possibility of peers. Such would be comic were it not the inevitable consequence of the constitution of the universe. To be shy of anything resembling himself is part and parcel of man's own individuality. Like the savage who fears nothing so much as a strange man, like Crusoe who grows pale at the sight of footprints not his own, the civilized thinker instinctively turns from the thought of mind other than the one he himself knows. To admit into his conception of the cosmos other finite minds as factors has in it something of the weird. Any hypothesis to explain the facts, no matter how improbable or even palpably absurd it be, is better than this. Snowcaps of solid carbonic acid gas, a planet cracked in a positively monomaniacal manner, meteors ploughing tracks across its surface with such mathematical precision that they must have been educated to the performance, and so forth and so on, in hypotheses each more astounding than its predecessor, commend themselves to man, if only by such means he may escape the admission of anything approaching his kind. Surely all this is puerile, and should be outgrown as speedily as possible. It is simply an instinct like any other, the projection of the instinct of self-preservation. We ought, therefore, to rise above it, and, where probably points to other things, boldly accept the fact provisionally, as we should the presence of oxygen, or iron, or anything else. Let us not cheat ourselves with words. Conservatism sounds finely, and covers any amount of ignorance and fear.

We must be just as careful not to run to the other extreme, and draw deductions of purely local outgrowth. To talk of Martian beings is not to mean Martian men. Just as the probabilities point to the one, so do they point away from the other. Even on this earth man is of the nature of an accident. He is the survival of by no means the highest physical organism. He is not even a high form of mammal. Mind has been his making. For aught we can see, some lizard or batrachian might just as well have popped into his place in the race, and been now the dominant creature of this earth. Under different physical circumstances he would have been certain to do so. Amid the physical surroundings that exist on Mars, we may be practically sure other organisms have been evolved which would strike us as exquisitely grotesque. What manner of beings they may be we have no data to conceive.

How diverse, however, they doubtless are from us will appear from such definite deduction as we are able to make from the physical differences between Mars and our earth. For example, the mere difference of gravity on the surface of the two planets is much more far-reaching in its effects than might at first be thought. Gravity on the surface of Mars is only a little more than one third what it is on the surface of the earth. This would work in two ways to very different conditions of existence from those to which we are accustomed. To begin with, three times as much work, as for example in digging a canal, could be done by the same expenditure of muscular force. If we were transported to Mars, we should be pleasingly surprised to find all our manual labor suddenly lightened threefold. But, indirectly, there might result a yet greater gain to our capabilities; for if Nature chose, she could afford there to build her inhabitants on three times the scale she does on earth, without their ever finding it out except by interplanetary comparison.

As we all know, a very large man is much more unwieldy than a very small one. An elephant refuses to hop like a flea; not because he considers it undignified to do so, but simply because he cannot take the step. If we could, we should all jump straight across the street, instead of painfully paddling through the mud. Our inability to do so depends partly on the size of the earth, and partly on the size of our own bodies, but not at all on what it at first seems entirely to depend on, the size of the street.

To see this, let us consider the very simplest case, that of standing erect. To this every-day feat opposes itself the weight of the body simply, a thing of three dimensions, height, breadth, and thickness, while the ability to accomplish it resides in the cross-section of the muscles of the knee, a thing of only two dimensions, breadth and thickness. Consequently, a person half as large again as another has about twice the supporting capacity of that other, but about three times as much to support. Standing therefore tires him out more quickly. If his size were to go on increasing, he would at last reach a stature at which he would no longer be able to stand at all, but would have to lie down. You shall see the same effect in quite inanimate objects. Take two cylinders of paraffine wax, one made into an ordinary candle, the other into a gigantic facsimile of one, and then stand both upon their bases. To the small one nothing happens. The big one, however, begins to settle, the base actually made viscous by the pressure of the weight above.

Now apply this principle to a possible inhabitant of Mars, and suppose him to be constructed three times as large as a human being in every dimension. If he were on earth, he would weigh twenty-seven times as much as the human being, but on the surface of Mars, since gravity there is only about one third of what it is here, he would weigh but nine times as much. The cross-section of his muscles would be nine times as great. Therefore the ratio of his supporting power to the weight he must support would be the same as ours. Consequently, he would be able to stand with no more fatigue than we experience. Now consider the work he might be able to do. His muscles, having length, breadth, and thickness, would all be twenty-seven times as effective as ours. He would prove twenty-seven times as strong as we, and could accomplish twenty-seven times as much. But he would further work upon what required, owing to decreased gravity, but one third the effort to overcome. His effective force, therefore, would be eighty-one times as great as man's, whether in digging canals or in other bodily occupation. As gravity on the surface of Mars is really a little more than one third that at the surface of the earth, the true ratio is not eighty-one, but about fifty; that is, a Martian would be, physically, fifty-fold more efficient than a man.

As the reader will observe, there is nothing problematical about this deduction whatever. It expresses an abstract ratio of physical capabilities which must exist between the two planets, quite irrespective of whether there be denizens on either, or how other conditions may further affect their forms.

Something more we may deduce about the characteristics of possible Martians, dependent upon Mars itself, a result of the age of the world they would live in.

A planet may in a very real sense be said to have a life of its own, of which what we call life may or may not be a detail. It is born, has its fiery youth, its sober middle age, its palsied senility, and ends at last in cold incapability of further change, its death. The speed with which it runs through its gamut of change depends upon its size; for the larger the body, the longer it takes to cool, and with it loss of heat means loss of life. It takes longer to cool because, as we saw in a previous paper, it has relatively more inside than outside, and it is through its outside that its inside cools. Now, inasmuch as time and space are not, as some philosophers have from their too mundane standpoint supposed, forms of our intellect, but essential attributes of the universe, the time taken by any process affects the character of the process itself, as does also the size of the body undergoing it. The changes brought about in a large planet by its cooling are not, therefore, the same as those brought about in a small one. Physically, chemically, and, to our present end, organically, the two results are quite diverse. So different, indeed, are they that unless the planet have at least a certain size it will never produce what we call life, meaning our particular chain of changes or closely allied forms of it, at all. As we saw in the case of atmosphere, it will lack even the premise to such conclusion.

Whatever the particular planet's line of development, however, in its own line it proceeds to greater and greater degrees of evolution, till the process is arrested by the planet's death, as above described. The point of development attained is, as regards its capabilities, precisely measured by the planet's own age, since the one is but a symptom of the other.

Now, in the special case of Mars, we have before us the spectacle of an old world, a world well on in years, a world much older relatively than the earth, halfway between it and the end we see so sadly typified by our moon, a body now practically past possibility of change. To so much

85

about his age Mars bears evidence on his face. He shows unmistakable signs of being old. What we know would follow advancing planetary years is legible there. His continents are all smoothed down; his oceans have all dried up. If he ever had a *jeunesse orageuse*, it has long since been forgotten. Although called after the turbulent of the gods, he is, and probably always has been, one of the most peaceful of the heavenly bodies. His name is a sad misnomer; indeed, the ancients seem to have been singularly unfortunate in their choice of planetary cognomens. With Mars so peaceful, Jupiter so young, and Venus bashfully dropped in cloud, the planets' names accord but ill with their temperaments.

Mars being thus old himself, we know that evolution on his surface must be similarly advanced. This only informs us of its condition relative to the planet's capabilities. Of its actual state our data are not definite enough to furnish much deduction. But from the fact that our own development has been comparatively a recent thing, and that a long time would be needed to bring even Mars to his present geological condition, we may judge any life he may support to be not only relatively, but really, more advanced than our own.

From the little we can see, such appears to be the case. The evidence of handicraft, if such it be, points to a highly intelligent mind behind it. Irrigation, unscientifically conducted, would not give us such truly wonderful mathematical fitness in the several parts to the whole as we there behold. A mind of no mean order would seem to have presided over the system we see,—a mind certainly of considerably more comprehensiveness than that which presides over the various departments of our own public works. Party politics, at all events, have had no part in them; for the system is planet wide. Quite possibly, such Martian folk are possessed of inventions of which we have not dreamed, and with them electrophones and kinetoscopes are things of a bygone past, preserved with veneration in museums as relics of the clumsy contrivances of the simple childhood of their kind. Certainly, what we see hints at the existence of beings who are in advance of, not behind us, in the race of life.

For answers to such problems we must look to the future. That Mars seems to be inhabited is not the last, but the first word on the subject. More important than the mere fact of the existence of living beings there is the question of what they may be like. Whether we ourselves shall live to learn this cannot, of course, be foretold. One thing, however, we can do, and that speedily: look at things from a standpoint raised above our local point of view; free our minds at least from the shackles that of necessity tether our bodies; recognize the possibility of others in the same light that we do the certainty of ourselves. That we are the sum and substance of the capabilities of the cosmos is something so preposterous as to be exquisitely comic. We pride ourselves upon being men of the world, forgetting that this is but objectionable singularity, unless we are in some wise men of more worlds than one. For after all, we are but a link in a chain. Man is merely this earth's highest production up to date. That he in any sense gauges the possibilities of the universe is humorous. He does not, as we can easily foresee, even gauge those of this planet. He has been steadily bettering from an immemorial past, and will apparently continue to improve through an incalculable future. Still less does he gauge the universe about him. He merely typifies in an imperfect way what is going on elsewhere, and what, to a mathematical certainty, is in some corners of the cosmos indefinitely excelled.

If astronomy teaches anything, it teaches that man is but a detail in the evolution of the universe, and that resemblant though diverse details are inevitably to be expected in the host of orbs around him. He learns that though he will probably never find his double anywhere, he is destined to discover any number of cousins scattered through space.

CPSIA information can be obtained
at www.ICGtesting.com
Printed in the USA
LVOW10*2115210817
545869LV00008B/27/P